ALGEBRA EXAMPLES

TRIGONOMETRY 2

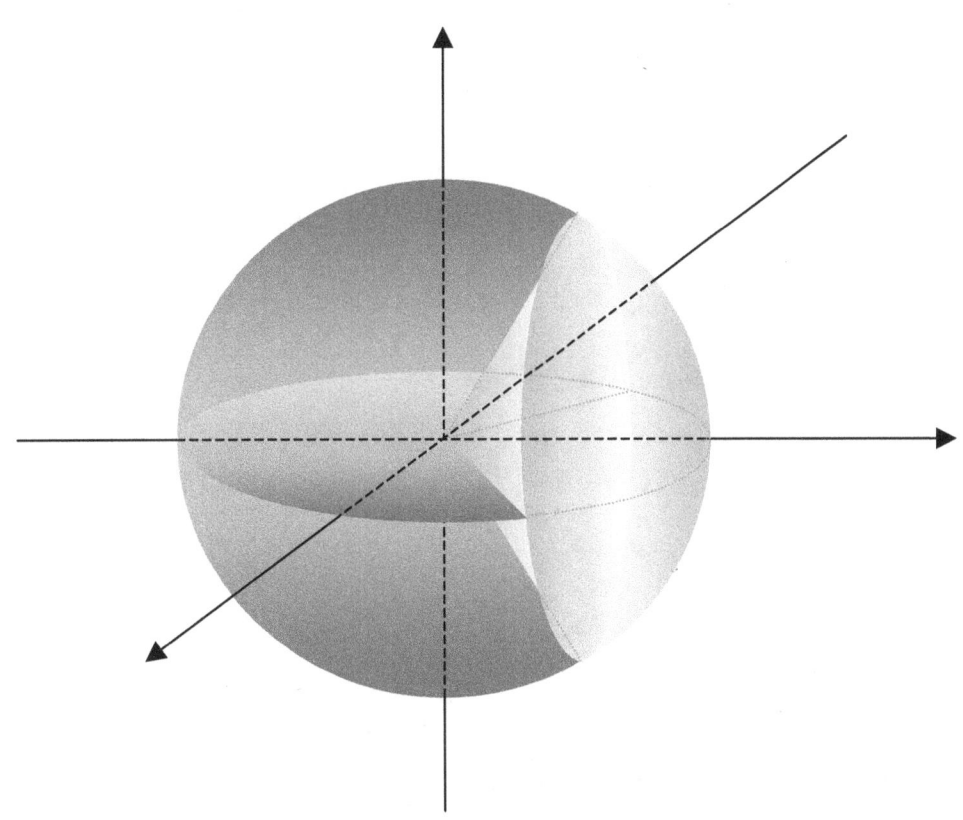

Seong R. KIM

Dear students:

Students need the best teacher, so you need examples, because examples are the best teacher. All the examples here are fully worked, and explain **how** the basic and essential tools in math are made, together with **what** they are, **how** they work, and **how** to work with them. Such tools include numbers, formulas, identities, equations, laws, etc.

Examples here begin with easy ones, of course. Covering every meter and yard properly, we can cover thousands of miles and kilometers. And it is particularly the case in math.

Of those examples therefore, some might even look too easy for you. It's not that easy though, to come up with those examples. Anyways, the bigger and the taller the tree, the deeper and the stronger the root.

Doing math, we work with ideas and run ideas, because every thing in math is an idea. A number is an idea, for instance, and the same is true for a line or circle, too. And putting ideas together, we build another, which becomes the base or an element of another, and each is connected. And that's the way your math grows. So you get to build a circuit, and sometimes, need to fill the gap or repair the circuit so that you get the sense of it.

So your calculation runs properly, and you get the problem solved.

The examples have been made and arranged so that they get tougher (or sometimes easier for some reason) as you proceed with them. In particular, similar examples with some variations are strategically repeated so that you can get the ideas or the tools tricky or complicated, and can get them mastered.

This book is however, nothing but a bunch of examples until you get it powered. How then, to get it powered, and make it run and work for you?

Just read it, and then, do each example in writing. And it is important to note that you do it in **your** writing. Just watching someone doing it, you just only feel that you can do it. If you do it, you can do it, but if you don't, we can hardly. It's a cliché, of course, but is always true that knowing is one thing and doing is another.

I've been helping students grow, take care of, and run their own math. The area covers algebra and geometry for high school or college students, and is especially for equations (for unknowns or curves), functions, and their graphs, which are the basic elements in calculus, which's been the core of my interest from my early age in high school.

Of my students, some are quite poor in math, and thus, are afraid of or hate math, some require special education because of exceptional intelligence, some are smart enough, some are naïve and diligent, some are clever but lazy, and most behave in general. All the students are badly after though, one thing in common: a strong and secure math skill. It is of course, the prime objective of my work, and I'm always happy to and eager to help them achieve it. The problem was however, that many of them wanted it to be purchased. And the question is, can we buy it?

We can buy the means, of course. And a solid math skill is feasible, too. We know however, we can't buy love, and the same is true for the math skill, too. It's not what we can buy or sell, and not what we can give or take. It is however, what we can grow, and need to grow. Your math grows as much as you grow and take care of it. So does mine.

What math then, do students most often do or use in high schools or colleges?

It is algebra and geometry. What algebra though?

Elementary algebra, of course
Doing the algebra, we work with numbers (many in kinds), constants, variables, ratios, rates, expressions, equations, inequalities, functions, identities, formulas, laws, etc., together with signs and symbols. And if we want to do algebra properly, we want to know their natures and how they mingle with each other.

So studying math ideas or tools, you want to know **what** they are, **how** they work, and **how** to work with them or **what** to do with them.   What then, about the geometry?

Basically, the geometry has much to do with shapes, positions, and angles. The shapes begin with triangles and circles, and move on to rectangles, squares, parallelograms or rhombuses, trapezoids, tetragons, other polygons, polyhedrons, etc.

Doing the geometry, too, though, we need to do the algebra stated above. So it is analytic geometry, often called coordinate geometry, too. And doing it, we can specify positions using coordinates. So in the geometry, basically, we work with graphs. Putting a math idea in a graph, we can not only effectively think about it but actually see it, too, and therefore, can efficiently work with it. What idea then, is it?

The idea begins with a point, line, parabola, circle, ellipse, and hyperbola, called a conic section or basic curve, and then, moves on to other curves, planes, surfaces, volumes, and other objects in various dimensional spaces, together with vectors.

And using an angle, we can specify an amount of turn or change in direction.

So learning, using, or applying those ideas or math tools, we get to solve problems.

And this book can help. It can help learn them, and use them so that you can navigate to find solutions to problems. And in particular, it can help come up with answers to those **what**s and **how**s stated above. So it can help you grow and run your own math, and thus, can help achieve your solid math skill.

It is however, not a magic book giving you a math skill of high caliber overnight. And it can have many mistakes, too. There is no magic, and math is full of facts and ideas. And it is after all, not me and not your teacher but you who put together some of those facts and ideas, and understand it. Putting facts and ideas together, understanding it, and taking care of what you have learned, you grow your math. And this book can help.

This is a book of examples designed to help you grow your math, and assumes that you are a real beginner. This book requires though, time and effort, the amount of which need to be substantial, too, but will be worth it. That's because you want a substantial achievement, and will get it. And probably, you will get to see this book helping you get there much faster than expected. And then, you will get to see the way math runs.

In math, everything is an idea. So is a problem. And solving it, we put it many different ways. For instance, while expanding or reducing it, or modifying or converting it, we keep searching for the solution, approaching the solution, and eventually, can get there. So don't look for the solution outside the problem. The solution is inside the problem if the problem is properly made.

If it is not, no solution is the solution. And in fact, it is often the case a problem itself is the solution. We can put a problem in many different ways, and eventually, can end up with the solution.   How come then, is the solution no other than the problem?

For instance, the solution to $3232 \div 101$ is 32.   And we can put it this way:

$$3232 \div 101 = \frac{3232}{101} = \frac{32 \times 101}{101} = \frac{32}{1} = 32 \Rightarrow 3232 \div 101 = 32.$$

And we can get this, too: $32 \Rightarrow 3232 \div 101$.   How?

$$32 = \frac{32}{1} = \frac{32 \times 101}{101} = \frac{3232}{101} = 3232/101 = 3232 \div 101.   \text{Too easy?}$$

For another instance, the solution to $ax^2 + bx + c = 0$ is: $x = \frac{-b \pm \sqrt{b^2 - 4ac}}{2a}$, which is called the quadratic formula.   How come then, is the solution no other than the problem?

We can put it this way:

$$x = \frac{-b \pm \sqrt{b^2 - 4ac}}{2a} \Rightarrow 2ax = -b \pm \sqrt{b^2 - 4ac} \Rightarrow 2ax + b = \pm\sqrt{b^2 - 4ac}$$

$$\Rightarrow (2ax + b)^2 = b^2 - 4ac \Rightarrow 4a^2x^2 + 4abx + b^2 = b^2 - 4ac$$

$$\Rightarrow 4a^2x^2 + 4abx = -4ac \Rightarrow ax^2 + bx = -c \Rightarrow ax^2 + bx + c = 0.$$

And we can get this, too: $ax^2 + bx + c = 0 \Rightarrow x = \frac{-b \pm \sqrt{b^2 - 4ac}}{2a}$.   How?

$$ax^2 + bx + c = a(x^2 + \tfrac{b}{a}x) + c = a(x^2 + \tfrac{b}{a}x + \tfrac{b^2}{4a^2} - \tfrac{b^2}{4a^2}) + c = a(x^2 + \tfrac{b}{a}x + \tfrac{b^2}{4a^2}) - \tfrac{b^2}{4a} + c$$

$$= a(x + \tfrac{b}{2a})^2 - \tfrac{b^2 - 4ac}{4a} = 0 \Rightarrow a(x + \tfrac{b}{2a})^2 = \tfrac{b^2 - 4ac}{4a} \Rightarrow (x + \tfrac{b}{2a})^2 = \tfrac{b^2 - 4ac}{4a^2} \Rightarrow x + \tfrac{b}{2a} = \pm\sqrt{\tfrac{b^2 - 4ac}{4a^2}}$$

$$\Rightarrow x = -\tfrac{b}{2a} \pm \tfrac{\sqrt{b^2 - 4ac}}{2a} = \tfrac{-b \pm \sqrt{b^2 - 4ac}}{2a} \Rightarrow x = \tfrac{-b \pm \sqrt{b^2 - 4ac}}{2a}.$$

And we call the set of processes above, algebra.

So if a problem is well defined, that is, if it makes sense, we should be able to get it solved the way below:

A problem ⇒ … ⇒ … ⇒ the solution, and thus: **the problem ⇒ the solution**.

So solving a problem, we put it many different ways so that we can get to the solution.

And that's the way, math runs.

May your math run very well.

Seong R. Kim

B.S. Math. Michigan Tech. Univ.　M.S. Math. Rensselaer Polytechnic Institute

Notes:

This book is about a math idea called trigonometry.

Why trigonometry though?

That's primarily because we often get to work with angles not only doing geometry but doing algebra, too. Doing high school math or college math, we can hardly avoid or stay away from algebra. And doing geometry, too, we often get to do algebra on expressions with angles. And doing such algebra, we can say that we do trig-algebra.

And next, we need to do trigonometry if we have to work with vectors and many other objects that have to do with angles. Working with such objects, we often need to find objects called components, which have directions or angles. Finding such components, we want to use some tools in trigonometry. And the tools are called trigonometric-ratios, often just called trig-ratios, for short.

So doing trigonometry, we get to use trig-ratios called sines, cosines, etc., together with a bit bigger tools called trig-identities, and some rules or formulas. So in this book, you get to know what those tools are and what they are about. That is to say that you will get to know how those tools work, what you can do with them, and how to work with them.

Specifically, you will get to learn, for instance, how the trig-ratio called the sine is made, what it is about, and how to use it. More specifically, you will see why you have to multiply by the sine, and what you get multiplying by it.

And of course, you will get to learn those important tools called trig-identities and the formulas called the Sine Rule and the Cosine Rule. Besides, you will get to see and will be familiar with some special tools called trig-functions.

And you will get them all through examples, that is, those tools will get explained with examples fully worked and detailed. Also, following steps to the solution in each example, you will be more familiar with the tools and the math ideas.

And you will get to strengthen your skill of algebra. So doing problems as well as learning ideas in math, you can do better and faster so that your math can run not only properly but fast enough, too.

And all the basics, tools, and ideas are covered in three books as well as in one book. And the three are as follows:

ALGEBRA EXAMPLES TRIGONOMETRY 1, which covers from the section Intro 1 to the section The Cosine Rule.

ALGEBRA EXAMPLES TRIGONOMETRY 2, which is this book, and covers from Examples 1 in The Cosine Rule to the section Sine Functions.

ALGEBRA EXAMPLES TRIGONOMETRY 3, which covers from Examples in Sine Functions to Examples in Trig-Algebra.

And all the contents of the three books above are put in one book as follows:

ALGEBRA EXAMPLES TRIGONOMETRY, which covers thus, from the section Intro 1 to Examples in Trig-Algebra.

Contents

In TRIGONOMETRY 2

The Preview of the Contents

In TRIGONOMETRY 3

The Preview of the Contents

In TRIGONOMETRY 1

$$(x + y)^2 = x^2 + 2xy + y^2. \qquad\qquad (x + y)^3 = x^3 + 3x^2y + 3xy^2 + y^3.$$

$$(x + y)(x - y) = x^2 - y^2. \qquad\qquad (x + y)(x^2 - xy + y^2) = x^3 + y^3.$$

$$(x^2 + xy + y^2)(x^2 - xy + y^2) = x^4 + x^2y^2 + y^4.$$

$$(x + a)(x + b) = x^2 + (a + b)x + ab. \qquad (ax + b)(cx + d) = acx^2 + (ad + bc)x + bd.$$

$$(x + a)(x + b)(x + c) = x^3 + (a + b + c)x^2 + (ac + bc + ca)x + abc.$$

$$(a + b + c)^2 = a^2 + b^2 + c^2 + 2(ab + bc + ca).$$

$$(a + b + c)(a^2 + b^2 + c^2 - ab - bc - ca) = a^3 + b^3 + c^3 - 3abc.$$

Suppose both a and $b \neq 0$, and both m and n are integers. Then, we get:

0. $a^m a^n = a^{m+n}$ **1.** $a^m / a^n = \dfrac{a^m}{a^n} = a^{m-n}$ **2.** $(a^m)^n = a^{mn}$

3. $(ab)^n = a^n b^n$ **4.** $(a/b)^n = \left(\dfrac{a}{b}\right)^n = a^n / b^n = \dfrac{a^n}{b^n}$

Suppose both a and $b > 0$, and m and n both are integers nonzero. Then, we get:

0.1. $a^{\frac{1}{n}} b^{\frac{1}{n}} = (ab)^{\frac{1}{n}}$.   **1.1.** $\dfrac{a^{\frac{1}{n}}}{b^{\frac{1}{n}}} = \left(\dfrac{a}{b}\right)^{\frac{1}{n}}$.   **2.1.** $(a^{\frac{1}{n}})^m = (a^m)^{\frac{1}{n}}$.

3.1. $(a^{\frac{1}{n}})^{\frac{1}{m}} = a^{\frac{1}{mn}} = (a^{\frac{1}{m}})^{\frac{1}{n}}$.   **3.2.** $(a^{mp})^{\frac{1}{np}} = (a^m)^{\frac{1}{n}}$, where p is a nonzero integer.

1. Suppose M, N, and $b > 0$, but $b \neq 1$, and we have: $A = \log_b M$, and $B = \log_b N$.
Then, we get: $A - B = \log_b M - \log_b N = \log_b \frac{M}{N}$.

2. Suppose that M and $b > 0$, but $b \neq 1$, and that we have: $E = \log_b M$.
Then, we get: $PE = P \log_b M = \log_b M^P$.

3. Suppose that a, b, C, and $D > 0$, but a and $b \neq 1$, and that we have: $\log_a C = \log_b D$.
Then, we get: $\log_a C = \log_b D = \log_{ab} CD$.

4. Suppose that a, b, C, and $D > 0$, but a and $b \neq 1$, and that we have: $\log_a C = \log_b D$.
Then, we get: $\log_a C = \log_b D = \log_{\frac{a}{b}} \frac{C}{D} = \log_{\frac{b}{a}} \frac{D}{C}$.

5. $\log_b b = 1$, and $\log_b 1 = 0$.   **6.** $\log_b A = \dfrac{\log_c A}{\log_c b}$.

7. $\log_b A = \dfrac{1}{\log_A b}$.

Note:

The drawings or graphs in this book are not exact, and are approximate or conceptual ones.

$\in$	"$a \in B$" means that a belongs to B. "$p, q,$ and $r \in W$" means that $p, q,$ and r belong to W.
$\Rightarrow$	"$A \Rightarrow B$." means that A implies B.
$\equiv$	$A \equiv B$ means that A and B are identical to each other.
$\neq$	$A \neq B$ means that A is not equal to B.
$\lvert A \rvert$	The magnitude of A. For instance, $\lvert -1 \rvert = \lvert 1 \rvert = 1$.
$\therefore$	Therefore
$\Leftrightarrow$	"$A \Leftrightarrow B$" means "If A then B." and "If B then A." We can read $A \Leftrightarrow B$ as "A if and only if B." In such a case, we can say that $A = B$.
Δx and Δy	Suppose that (x_1, y_1) and (x_2, y_2) are two points in the x-y plane. Then, we get either of the two below. $\Delta x = x_2 - x_1$, and $\Delta y = y_2 - y_1$. $\Delta x = x_1 - x_2$, and $\Delta y = y_1 - y_2$.

Distance Formula

Suppose that d is the distance between two points (x_1, y_1) and (x_2, y_2) in the x-y plane. Then, we get $d^2 = (\Delta x)^2 + (\Delta y)^2$.

Examples 1 in The Cosine Rule

Doing all these examples, refer to the triangle **ABC** below:

Fig. 0

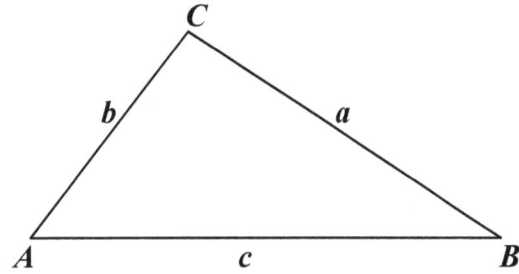

0. Show that: $0 < A < \pi/2 \Rightarrow a^2 < b^2 + c^2$.

1. Show that: $\pi/2 < A < \pi \Rightarrow a^2 > b^2 + c^2$

2. Assuming: $A = \pi/3$, $b = 40$, and $c = 20(\sqrt{3} + 1)$, find a, B, and C.

3. Assuming: $a = \sqrt{6}$, $b = 2\sqrt{3}$, and $c = 3 + \sqrt{3}$, find A, B, and C.

Suggestions or Solutions
To the Problem in the Example 0

Show that: $0 < A < \pi/2 \Rightarrow a^2 < b^2 + c^2$ for the triangle ABC below:

Fig. 0.0

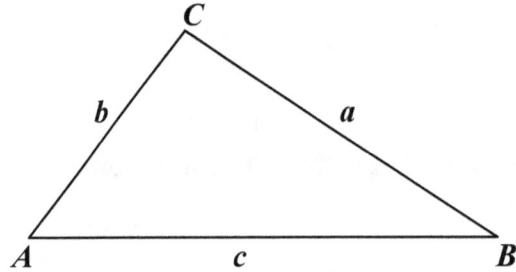

We have: $a^2 = b^2 + c^2 - 2bc \cos A$.

And we have: $bc > 0$, and $0 < A < \pi/2 \Rightarrow \cos A > 0$.

So we get: $b^2 + c^2 - a^2 = 2bc \cos A > 0 \Rightarrow b^2 + c^2 - a^2 > 0 \Rightarrow b^2 + c^2 > a^2$.

And thus, we get: $0 < A < \pi/2 \Rightarrow a^2 < b^2 + c^2$.

If not quite sure of the idea behind the processes above, follow the steps below:

The relation given looks like the triangle inequality, and the inequality says:
$a + b > c$, $b + c > a$, or $c + b > a$.

The inequality does not though, explain the relation given.

We have however, a trig-identity that shows a relation between the squares of the sides of a triangle. What then, is the trig-identity?

It is the cosine rule, where:

$a^2 = b^2 + c^2 - 2bc \cos A$, $b^2 = c^2 + a^2 - 2ca \cos B$, or $c^2 = a^2 + b^2 - 2ab \cos C$.

So using the cosine rule, along with $0 < A < \pi/2$, we can begin with an equation below:

$a^2 = b^2 + c^2 - 2bc \cos A$.

And next, we can put it this way: $b^2 + c^2 - a^2 = 2bc \cos A$.

And we know that $0 < A < \pi/2 \Rightarrow \cos A > 0$, and that $bc > 0$ since $b > 0$ and $c > 0$.

So we get: $b^2 + c^2 - a^2 = 2bc \cos A > 0$.

That is to say that we can get: $b^2 + c^2 - a^2 > 0$. And thus, we get: $b^2 + c^2 > a^2$.

So we get: $0 < A < \pi/2 \Rightarrow a^2 < b^2 + c^2$.

In short:

We have: $a^2 = b^2 + c^2 - 2bc \cos A$.

And we have: $bc > 0$, and $0 < A < \pi/2 \Rightarrow \cos A > 0$.

So we get: $b^2 + c^2 - a^2 = 2bc \cos A > 0 \Rightarrow b^2 + c^2 - a^2 > 0 \Rightarrow b^2 + c^2 > a^2$.

And thus, we get: $0 < A < \pi/2 \Rightarrow a^2 < b^2 + c^2$.

Suggestions or Solutions
To the **Problem** in the Example **1**

Show that: $\pi/2 < A < \pi \Rightarrow a^2 > b^2 + c^2$ for the triangle ABC below:

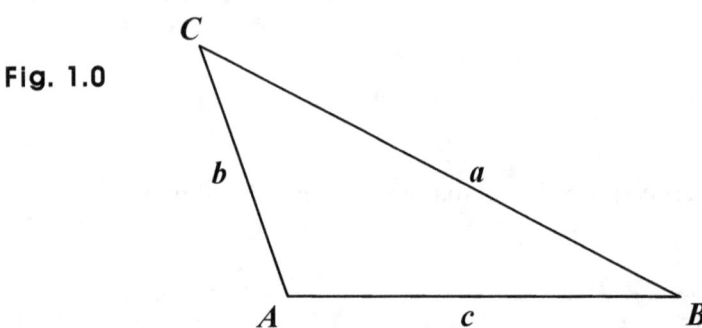

Fig. 1.0

We have: $a^2 = b^2 + c^2 - 2bc \cos A$.

And we have: $bc > 0$, and $\pi/2 < A < \pi \Rightarrow \cos A < 0$.

So we get: $b^2 + c^2 - a^2 = 2bc \cos A < 0 \Rightarrow b^2 + c^2 - a^2 < 0 \Rightarrow b^2 + c^2 < a^2$.

And thus, we get: $\pi/2 < A < \pi \Rightarrow a^2 > b^2 + c^2$.

If not quite sure of the idea behind the processes above, follow the steps below:

This example is no other than the example 0 covered earlier.

So we can use again, the cosine rule, where:

$a^2 = b^2 + c^2 - 2bc \cos A$, $b^2 = c^2 + a^2 - 2ca \cos B$, or $c^2 = a^2 + b^2 - 2ab \cos C$.

Using thus, the cosine rule, along with $0 < A < \pi/2$, we can begin with an equation as follows: $a^2 = b^2 + c^2 - 2bc \cos A$.

And next, we can put it this way again: $b^2 + c^2 - a^2 = 2bc \cos A$. What then?

We know that $\pi/2 < A < \pi \Rightarrow \cos A < 0$, and that $bc > 0$.

So we get: $b^2 + c^2 - a^2 = 2bc \cos A < 0$.

That is, we can get: $b^2 + c^2 - a^2 < 0$. And thus, we get: $b^2 + c^2 < a^2$.

So we get: $\pi/2 < A < \pi \Rightarrow a^2 > b^2 + c^2$.

In short:

We have: $a^2 = b^2 + c^2 - 2bc \cos A$.

And we have: $bc > 0$, and $\pi/2 < A < \pi \Rightarrow \cos A < 0$.

So we get: $b^2 + c^2 - a^2 = 2bc \cos A < 0 \Rightarrow b^2 + c^2 - a^2 < 0 \Rightarrow b^2 + c^2 < a^2$.

And thus, we get: $\pi/2 < A < \pi \Rightarrow a^2 > b^2 + c^2$.

Suggestions or Solutions

To the **Problem** in the Example **2**

Find *a*, *B*, and *C* assuming: $A = \pi/3$, $b = 40$, and $c = 20(\sqrt{3} + 1)$ for the triangle *ABC* below:

Fig. 2.0

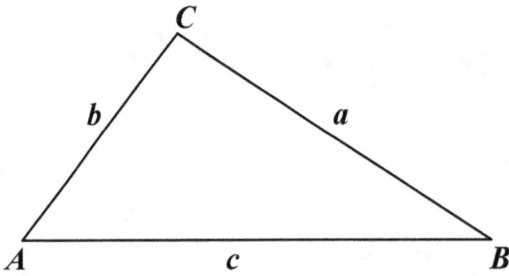

$a^2 = b^2 + c^2 - 2bc \cos A \Rightarrow a^2 = 40^2 + 20^2(\sqrt{3} + 1)^2 - 2 \cdot 40 \cdot 20(\sqrt{3} + 1) \cos \pi/3$

$= 1600 + 400(3 + 2\sqrt{3} + 1) - 1600(\sqrt{3} + 1) \cdot (1/2) = 2400 \Rightarrow a^2 = 2400 \Rightarrow a = \pm 20\sqrt{6}$.

So we get: $a = 20\sqrt{6}$ since $a > 0$.

And next, using the sine rule, we get: $\dfrac{a}{\sin A} = \dfrac{b}{\sin B} \Rightarrow \dfrac{20\sqrt{6}}{\sqrt{3}/2} = 40\sqrt{2} = \dfrac{40}{\sin B}$.

So we get: $\sin B = \dfrac{1}{\sqrt{2}}$. And thus, we get: $B = \pi/4$ or $\pi - \pi/4 = 3\pi/4$.

Assuming however, $B = \pi - \pi/4 = 3\pi/4$, we get: $A + B = \pi/3 + 3\pi/4 = 13\pi/12 > \pi$, which is not possible, because $A + B + C = \pi$, and $C > 0$.

So we get: $B \neq 3\pi/4$, and $B = \pi/4$.

And we have: $A + B + C = \pi$.

Thus, we get: $C = \pi - A - B = \pi - \pi/3 - \pi/4 = 5\pi/12$.

If not quite sure of the idea behind the processes above, follow the steps below:

To begin with, knowing two angles in a triangle, we can get the other angle, since the sum of all the three angles is π.

In this case however, we know one angle only, which is A.
So how can we find the other two angles, B and C?

We know the angle A, and two sides, b and c. And we have the sine rule below:

$$\frac{a}{\sin A} = \frac{b}{\sin B} = \frac{c}{\sin C} = 2R. \quad \text{So?}$$

So finding the side a, we can get the angles B and C both.
How then, can we get the side a?

We know that the angle A is between the two sides b and c.
So what can we use to find the side a?
We have the cosine rule, where $a^2 = b^2 + c^2 - 2bc \cos A$.

So using the cosine rule, we can get the side a. So using the rule, we get:

$$a^2 = b^2 + c^2 - 2bc \cos A \Rightarrow a^2 = 40^2 + 20^2(\sqrt{3}+1)^2 - 2 \cdot 40 \cdot 20(\sqrt{3}+1) \cos \pi/3$$

$$= 1600 + 400(3 + 2\sqrt{3} + 1) - 1600(\sqrt{3}+1) \cdot (1/2) = 1600 + 1600 - 800 = 2400$$

$$\Rightarrow a^2 = 2400 \Rightarrow a = \pm 20\sqrt{6}.$$

So we get: $a = 20\sqrt{6}$, since $a > 0$. What then, can we do with the value of a?

We can get the value of the angle B using the sine rule.

8

Thus next, using the sine rule, we get: $\dfrac{a}{\sin A} = \dfrac{b}{\sin B} \Rightarrow \dfrac{20\sqrt{6}}{\sqrt{3}/2} = \dfrac{40}{\sin B}$.

And we have: $\dfrac{20\sqrt{6}}{\sqrt{3}/2} = \dfrac{2\cdot 20\sqrt{2}\sqrt{3}}{\sqrt{3}} = 40\sqrt{2}$.

So we get: $\dfrac{40}{\sin B} = 40\sqrt{2} \Rightarrow \sin B = \dfrac{1}{\sqrt{2}}$.

And thus, we get: $B = \pi/4$ or $\pi - \pi/4$, since we have: $\sin(\pi - \theta) = \sin\theta$ for $0 \le \theta \le \pi/2$.

Assuming however, $B = \pi - \pi/4 = 3\pi/4$, we get: $A + B = \pi/3 + 3\pi/4 = 13\pi/12 > \pi$, which is not possible, since A and B are two angles in the triangle ABC, and the sum of the three angles in a triangle is π, that is, $A + B + C = \pi$.

So we get: $B \ne 3\pi/4$, and $B = \pi/4$.

And thus, since $B = \pi/4$, and $A = \pi/3$, we get: $C = \pi - A - B = \pi - \pi/3 - \pi/4 = 5\pi/12$.

Suggestions or Solutions
To the **Problem** in the Example **3**

Find *A*, *B*, and *C* assuming: $a = \sqrt{6}, b = 2\sqrt{3}$, and $c = 3 + \sqrt{3}$ for the triangle below:

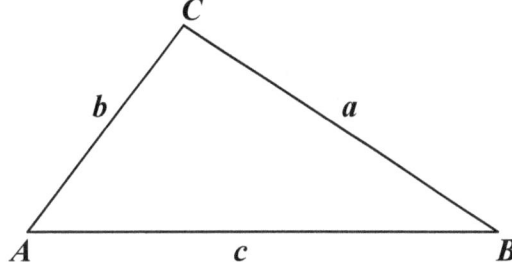

Fig. 3.0

We have all the three sides in a triangle, but have none of the three angles.

So what can we use to find all the three angles?

We have a formula where an angle in triangle can be expressed in terms of the sides.

What then, is the formula?

The formula is in fact, called a rule, and is known as the cosine rule. And the rule says:

$a^2 = b^2 + c^2 - 2bc \cos A$, $b^2 = c^2 + a^2 - 2ca \cos B$, and $c^2 = a^2 + b^2 - 2ab \cos C$.

Knowing thus, the three sides, we can get all the three angles.

In other words, we can get:

$$a^2 = b^2 + c^2 - 2bc \cos A \Rightarrow \cos A = \frac{b^2 + c^2 - a^2}{2bc}.$$

$$b^2 = c^2 + a^2 - 2ca \cos B \Rightarrow \cos B = \frac{c^2 + a^2 - b^2}{2ca}.$$

$$c^2 = a^2 + b^2 - 2ab \cos C \Rightarrow \cos C = \frac{a^2 + b^2 - c^2}{2ab}.$$

And thus, beginning with the angle A, we get:

$$a^2 = b^2 + c^2 - 2bc \cos A \Rightarrow 6 = 12 + (3+\sqrt{3})^2 - 2 \cdot 2\sqrt{3}(3+\sqrt{3}) \cos A$$

$$= 12 + 9 + 6\sqrt{3} + 3 - (12\sqrt{3} + 12) \cos A = 24 + 6\sqrt{3} - 12(\sqrt{3}+1) \cos A$$

$$\Rightarrow \cos A = \frac{18 + 6\sqrt{3}}{12(\sqrt{3}+1)} = \frac{6(3+\sqrt{3})}{12(\sqrt{3}+1)} = \frac{\sqrt{3}(\sqrt{3}+1)}{2(\sqrt{3}+1)} = \frac{\sqrt{3}}{2} \Rightarrow \cos A = \frac{\sqrt{3}}{2}.$$

So we get: $A = \pi/6$.

What then, is the next?

Next, using the cosine rule again, we get:

$$b^2 = c^2 + a^2 - 2ca \cos B \Rightarrow 12 = (3+\sqrt{3})^2 + 6 - 2 \cdot \sqrt{6}(3+\sqrt{3}) \cos B$$

$$= 6 + 9 + 6\sqrt{3} + 3 - 2(3\sqrt{6} + 3\sqrt{2}) \cos B = 18 + 6\sqrt{3} - 6\sqrt{2}(\sqrt{3}+1) \cos B$$

$$\Rightarrow \cos B = \frac{6 + 6\sqrt{3}}{6\sqrt{2}(\sqrt{3}+1)} = \frac{6(\sqrt{3}+1)}{6\sqrt{2}(\sqrt{3}+1)} = \frac{1}{\sqrt{2}} \Rightarrow \cos B = \frac{1}{\sqrt{2}}.$$

So we get: $B = \pi/4$. How then, can we get C?

We have: $A + B + C = \pi$, since A, B, and C are the three angles in the triangle ABC, and the sum of the three angles in a triangle is π.

So we get: $C = \pi - A - B = \pi - \pi/6 - \pi/4 = 7\pi/12$.

Examples 2 in The Cosine Rule

Doing all these examples, refer to the triangle ABC below:

Fig. 0

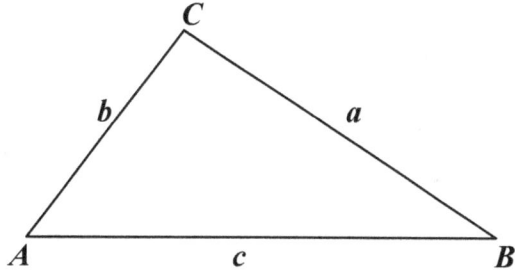

0.   Assuming $a = x^2 + x + 1$, $b = x^2 - 1$, and $c = 2x + 1$ for x real, determine the value of x so that a, b, and c are the three sides of the triangle ABC.

1.   Assuming $a = x^2 + x + 1$, $b = x^2 - 1$, and $c = 2x + 1$ for x real, and a, b, and c are the three sides of the triangle ABC, find the biggest of the three angles in the triangle ABC.

2.   Find the kind of the triangle ABC that satisfies an equation: $\sin A = 2(\cos B)(\sin C)$.

3.   Find the kind of the triangle ABC that satisfies an equation: $a \cos A = b \cos B$.

4.   Find the kind of the triangle ABC that satisfies: $(b - c) \cos^2 A = b \cos^2 B - c \cos^2 C$.

Suggestions or Solutions
To the **Problem** in the Example **0**

Assuming $a = x^2 + x + 1$, $b = x^2 - 1$, **and** $c = 2x + 1$ **for** x **real, determine the value of** x **so that** a, b, **and** c **are the three sides of the triangle** ABC **below:**

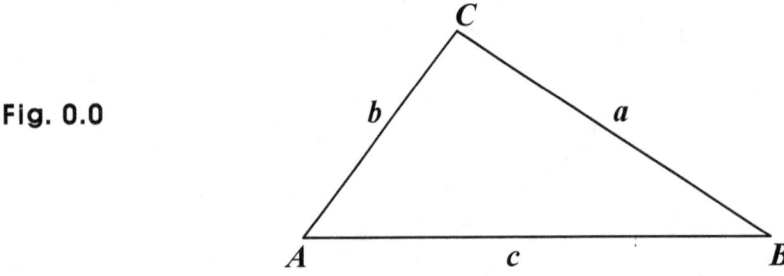

Fig. 0.0

Forming a triangle with three line segments, we cannot just use any line segments. What line segments then, can we use forming a triangle?

In other words, what is the condition on the line segments a, b, and c if they can form a triangle?

The three line segments have to meet the triangle inequality, which says:

$a + b > c$, $b + c > a$, or $c + a > b$.   Which one of the three though, we want to apply?

Assuming a is the largest of all the three sides, we get: $b + c > a$, which is the one to be applied. That is to say that if it turns out that a is the largest, we want to apply the inequality above. Is the equality all then, that we need to apply?

We need to have: $a > 0$, $b > 0$, and $c > 0$, of course, which is natural, since sides are positive. And thus, the set of three inequalities is the basic condition that the three line segments a, b, and c have to meet. Let's now, begin with the basic condition.

What then, do we need to begin with?

To begin with, we have: $a = x^2 + x + 1$, $b = x^2 - 1$, and $c = 2x + 1$ for x real.

So first, beginning with the basic condition stated above, we get:

$a = x^2 + x + 1 = (x + 1/2)^2 + 3/4 > 0$ for all x.

$b = x^2 - 1 = (x - 1)(x + 1) > 0 \Rightarrow x > 1$ or $x < -1$.

$c = 2x + 1 > 0 \Rightarrow x > -1/2$.

And thus, x has to meet, at the same time, all the three cases below:
One is that x is real. Another is that $x > 1$ or $x < -1$. And the other is that $x > -1/2$.

Then, getting the extent satisfying all three at the same time, we get: $x > 1$.
What then, is the next?

We have to meet the triangle inequality. So first, finding the largest side, we get:

$a - b = x^2 + x + 1 - (x^2 - 1) = x + 2 > 0$ since $x > 1$, and thus: $a > b$.

$a - c = x^2 + x + 1 - (2x + 1) = x^2 - x = x(x - 1) > 0$ since $x > 1$, and thus, $a > c$.

So a is the largest. And thus, we want to apply this inequality: $b + c > a$.
And applying the triangle inequality, we can get first: $b + c > a \Rightarrow b + c - a > 0$.

And we have: $a = x^2 + x + 1$, $b = x^2 - 1$, and $c = 2x + 1$.

So we get: $b + c - a = (x^2 - 1) + (2x + 1) - (x^2 + x + 1) = x - 1 > 0 \Rightarrow x > 1$.

And also, we get: $x > 1$ from the basic condition, too.

Thus, putting threads together, we get: $x > 1$.

Suggestions or Solutions
To the **Problem** in the Example **1**

Assuming $a = x^2 + x + 1$, $b = x^2 - 1$, and $c = 2x + 1$ for $x > 1$, and a, b, and c are the three sides of the triangle ABC below, find the biggest of the three angles in the triangle ABC.

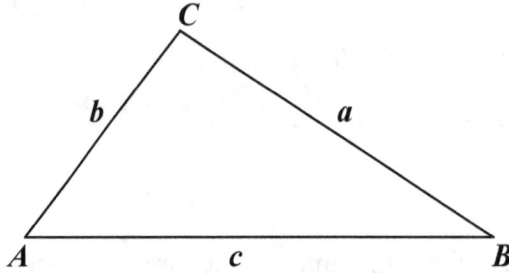

Fig. 1.0

To begin with, we get: $a - b = x^2 + x + 1 - (x^2 - 1) = x + 2 > 0$ for $x > 1$, and also,

$a - c = x^2 + x + 1 - (2x + 1) = x^2 - x = x(x - 1) > 0$ for $x > 1$.

So the angle A is the biggest of the three angles, since it faces the largest side, which is a.

Next, we have: $a^2 = b^2 + c^2 - 2bc \cos A$.

So we get: $\cos A = \dfrac{b^2 + c^2 - a^2}{2bc} \Rightarrow \cos A = \dfrac{(x^2 - 1)^2 + (2x + 1)^2 - (x^2 + x + 1)^2}{2(x^2 - 1)(2x + 1)}$.

Meanwhile, we can have: $(x^2 - 1)^2 + (2x + 1)^2 - (x^2 + x + 1)^2$

$= x^4 - 2x^2 + 1 + 4x^2 + 4x + 1 - (x^4 + x^2 + 1 + 2x^3 + 2x + 2x^2)$

$= 2x + 1 - (x^2 + 2x^3) = -2x^3 - x^2 + 2x + 1$, and also, we can get:

$2(x^2 - 1)(2x + 1) = 2(2x^3 + x^2 - 2x - 1)$.

So we get: $\cos A = \dfrac{-2x^3 - x^2 + 2x + 1}{2(2x^3 + x^2 - 2x - 1)} = \dfrac{-(2x^3 + x^2 - 2x - 1)}{2(2x^3 + x^2 - 2x - 1)} = -\dfrac{1}{2}$.

And we have: $\cos \pi/3 = 1/2$.   So we get: $-\cos \pi/3 = -1/2$.

And we have: $\cos (\pi - \theta) = -\cos \theta$.   And thus, we get: $A = \pi - \pi/3 = 2\pi/3$.

If not quite sure of the idea behind the processes above, follow the steps below:

To begin with, we have a fact that in the triangle, the longest side faces the largest angle of the three angles.

For instance, assuming A is the largest angle, we can say that the triangle is as follows:

Fig. 1.1

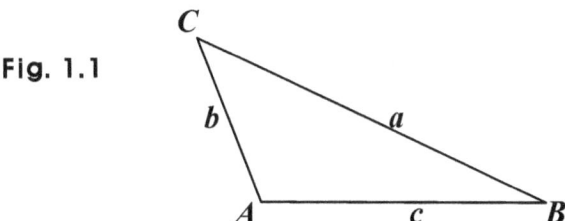

And in fact, we get:

$a - b = x^2 + x + 1 - (x^2 - 1) = x + 2 > 0$ for $x > 1$, and also,

$a - c = x^2 + x + 1 - (2x + 1) = x^2 - x = x(x - 1) > 0$ for $x > 1$.

So we can see A is the biggest of the three angles, and faces the largest side, which is a.

And in fact, the larger the angle is, the larger the facing side is.

How can we get the angle A though?

We have the cosine rule, where: $a^2 = b^2 + c^2 - 2bc \cos A$.

And we have this, too: $a = x^2 + x + 1$, $b = x^2 - 1$, and $c = 2x + 1$ for $x > 1$.

So applying the cosine rule, we get first: $\cos A = \dfrac{b^2 + c^2 - a^2}{2bc}$, and thus, can set:

$$\cos A = \frac{(x^2 - 1)^2 + (2x + 1)^2 - (x^2 + x + 1)^2}{2(x^2 - 1)(2x + 1)}.$$

Meanwhile, we can have:

$$(x^2 - 1)^2 + (2x + 1)^2 - (x^2 + x + 1)^2$$

$$= x^4 - 2x^2 + 1 + 4x^2 + 4x + 1 - (x^4 + x^2 + 1 + 2x^3 + 2x + 2x^2)$$

$$= 2x + 1 - (x^2 + 2x^3) = -2x^3 - x^2 + 2x + 1.$$

And also, we can get: $2(x^2 - 1)(2x + 1) = 2(2x^3 + x^2 - 2x - 1)$.

So we get: $\cos A = \dfrac{-2x^3 - x^2 + 2x + 1}{2(2x^3 + x^2 - 2x - 1)} = \dfrac{-(2x^3 + x^2 - 2x - 1)}{2(2x^3 + x^2 - 2x - 1)} = -\dfrac{1}{2}.$

What then, is the angle A?

We have a trig-identity where: $\cos(\pi - \theta) = -\cos\theta$.

And we have: $\cos \pi/3 = 1/2$, too.

So we get: $-\cos \pi/3 = -1/2$.

And thus, we get: $A = \pi - \pi/3 = 2\pi/3$.

Suggestions or Solutions
To the **Problem** in the Example **2**

Find the kind of the triangle ABC that satisfies an equation: $\sin A = 2(\cos B)(\sin C)$.

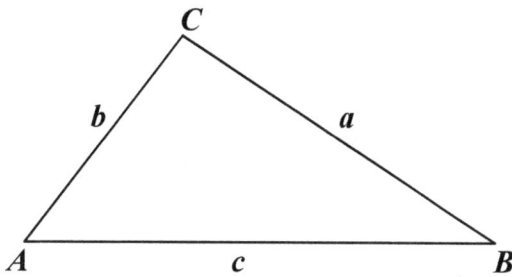

Fig. 2.0

We have: $\dfrac{a}{\sin A} = \dfrac{b}{\sin B} = \dfrac{c}{\sin C} = 2R$, and $b^2 = c^2 + a^2 - 2ca \cos B$.

So first, we can get: $\sin A = a/(2R)$, $\cos B = \dfrac{c^2 + a^2 - b^2}{2ca}$, and $\sin C = (c/2R)$.

So next, putting the above into the equation where $\sin A = 2(\cos B)(\sin C)$, we get:

$$\frac{a}{2R} = \frac{2(c^2 + a^2 - b^2)}{2ca} \cdot \frac{c}{2R} \Rightarrow a = \frac{(c^2 + a^2 - b^2)}{a}.$$

Thus, we get: $a^2 = c^2 + a^2 - b^2 \Rightarrow c^2 = b^2 \Rightarrow c = b$.

Assuming now, $a = b = c$, we get: $\cos B = \dfrac{c^2 + a^2 - b^2}{2ca} = \dfrac{a^2 + a^2 - a^2}{2aa} = \dfrac{1}{2}$.

So we get: $B = \pi/3$. And next, we have: $\dfrac{a}{\sin A} = \dfrac{b}{\sin B} = \dfrac{c}{\sin C} = 2R$.

Setting thus, $a = b = c$, we get: $\sin A = \sin B = \sin C$.   So we get: $A = C = \pi/3$, too.

And thus, it is the case where $a = b = c$.   So the triangle ABC is a regular triangle.

If not quite sure of the idea behind the processes above, follow the steps below:

How do we know if a triangle is in a particular kind?

We could see it finding the relation among the sides. How then, can we find the relation?

We are given an equation made of **sin** A, **cos** B, and **sin** C. So what?

We can put the equation in terms of the three sides, a, b, and c.

That is to say that we can get a connective expression between the three sides.
And the expression can show the relation. How then, can we get the expression?

The sine rule has: **sin** A and **sin** C, together with **sin** B.
And the cosine rule has: **cos** B, together with the three sides, a, b, and c.

The sine rule is: $\dfrac{a}{\sin A} = \dfrac{b}{\sin B} = \dfrac{c}{\sin C} = 2R.$

And the cosine rule is: $b^2 = c^2 + a^2 - 2ca\,\cos B$.
So getting the expression we need, we can use the sine rule and the cosine rule. How?

Using the sine rule first, we can get: **sin** $A = a/(2R)$, and **sin** $C = (c/2R)$.

And next, using the cosine rule, we can get: $\cos B = \dfrac{c^2 + a^2 - b^2}{2ca}.$

So next, putting the expressions above into the equation where **sin** $A = 2(\cos B)(\sin C)$,

we get: $\dfrac{a}{2R} = \dfrac{2(c^2 + a^2 - b^2)}{2ca} \cdot \dfrac{c}{2R} \Rightarrow a = \dfrac{(c^2 + a^2 - b^2)}{a}.$

Thus, we get: $a^2 = c^2 + a^2 - b^2 \Rightarrow c^2 = b^2 \Rightarrow c = b$.   What triangle then, is it?

It is a triangle where two sides are equal, and thus, is an isosceles triangle.

Can it not be though, a regular triangle?

Assuming now, $a = b = c$, we get: $\cos B = \dfrac{c^2 + a^2 - b^2}{2ca} = \dfrac{a^2 + a^2 - a^2}{2aa} = \dfrac{1}{2}$.

So we get: $B = \pi/3$, which is $60°$, which indicates it can be a regular triangle.

And next, we have: $\dfrac{a}{\sin A} = \dfrac{b}{\sin B} = \dfrac{c}{\sin C} = 2R$.

Setting thus, $a = b = c$, we get: $\sin A = \sin B = \sin C$.

We know $B = \pi/3$.   So we get: $A = C = \pi/3$, too.

And thus, it is the case where $a = b = c$.   So the triangle ABC is a regular triangle.

Note that a regular triangle is an isosceles triangle because two sides are equal in length in a regular triangle. That is, all regular triangles can be said to be isosceles. It is the case however, an isosceles triangle may not be a regular triangle.

Suggestions or Solutions
To the Problem in the Example 3

Find the kind of the triangle *ABC* that satisfies an equation: *a* cos *A* = *b* cos *B*.

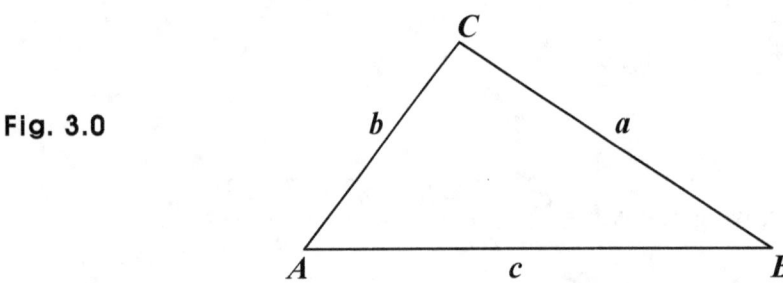

Fig. 3.0

We have the cosine rule, where: $a^2 = b^2 + c^2 - 2bc \cos A$, and $b^2 = c^2 + a^2 - 2ca \cos B$.

So we get: $a \cos A = \dfrac{a(b^2 + c^2 - a^2)}{2bc}$, and $b \cos B = \dfrac{b(c^2 + a^2 - b^2)}{2ca}$.

And we have: ***a* cos *A* = *b* cos *B*.** Thus, we get:

$$\frac{a(b^2 + c^2 - a^2)}{2bc} = \frac{b(c^2 + a^2 - b^2)}{2ca} \Rightarrow \frac{a(b^2 + c^2 - a^2)}{b} = \frac{b(c^2 + a^2 - b^2)}{a}.$$

So multiplying both sides by *ab*, we get:

$a^2(b^2 + c^2 - a^2) = b^2(c^2 + a^2 - b^2)$

$\Rightarrow a^2(b^2 + c^2 - a^2 - 2b^2 + 2b^2) = b^2(c^2 + a^2 - b^2 - 2a^2 + 2a^2)$

$\Rightarrow a^2(c^2 - a^2 - b^2) + 2a^2b^2 = b^2(c^2 - a^2 - b^2) + 2a^2b^2$

$\Rightarrow a^2(c^2 - a^2 - b^2) = b^2(c^2 - a^2 - b^2)$

$\Rightarrow (a^2 - b^2)(c^2 - a^2 - b^2) = 0 \Rightarrow a^2 = b^2$ or $c^2 = a^2 + b^2$.

And therefore, it can be any of the three cases below:

- It is an isosceles triangle with $a = b$.

- It is a right triangle with $C = \pi/2$.

- It is an isosceles right triangle with $C = \pi/2$.

And of course, we can do the algebra the way below, too:

$$a^2(b^2 + c^2 - a^2) = b^2(c^2 + a^2 - b^2)$$

$$\Rightarrow a^2b^2 + a^2c^2 - a^4 = b^2c^2 + b^2a^2 - b^4 \Rightarrow a^2c^2 - a^4 = b^2c^2 - b^4$$

$$\Rightarrow a^2c^2 - a^4 - b^2c^2 + b^4 = c^2(a^2 - b^2) - (a^4 - b^4) = c^2(a^2 - b^2) - (a^2 - b^2)(a^2 + b^2)$$

$$= (a^2 - b^2)(c^2 - a^2 - b^2) = 0 \Rightarrow a^2 = b^2 \text{ or } c^2 = a^2 + b^2.$$

Suggestions or Solutions
To the **Problem** in the Example **4**

Find the kind of the triangle ABC that satisfies: $(b-c)\cos^2 A = b\cos^2 B - c\cos^2 C$.

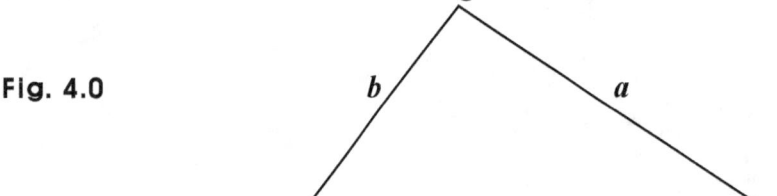

Fig. 4.0

The equation given is put in terms of the sides and the cosines of the three angles.
So it looks like we want to try the cosine rule.

The algebra gets easier though, applying the sine rule first.
How then, can we apply the sine rule?

We have a trig-identity where $\sin^2\theta + \cos^2\theta = 1$.

So applying first, the identity above, we can get: $(b-c)\cos^2 A = b\cos^2 B - c\cos^2 C$
$\Rightarrow (b-c)(1-\sin^2 A) = b(1-\sin^2 B) - c(1-\sin^2 C)$.

Meanwhile, we can have: $(b-c)(1-\sin^2 A) = (b-c) - (b-c)\sin^2 A$, and
$b(1-\sin^2 B) - c(1-\sin^2 C) = (b-c) - b\sin^2 B + c\sin^2 C$.

So we get: $(b-c) - (b-c)\sin^2 A = (b-c) - b\sin^2 B + c\sin^2 C$
$\Rightarrow -(b-c)\sin^2 A = -b\sin^2 B + c\sin^2 C$.

That is, we get: $(b-c)\sin^2 A = b\sin^2 B - c\sin^2 C$.   What then, is the next?

We have the sine rule, where: $\dfrac{a}{\sin A} = \dfrac{b}{\sin B} = \dfrac{c}{\sin C} = 2R$.

That is, we have: $\sin A = a/(2R)$, $\sin B = b/(2R)$, and $\sin C = c/(2R)$.

So we get: $(b - c)a^2/(4R^2) = b \cdot b^2/(4R^2) - c \cdot c^2/(4R^2)$. That is, we get: $(b - c)a^2 = b^3 - c^3$.

And we have: $b^3 - c^3 = (b - c)(b^2 + bc + c^2)$.

So we get: $(b - c)a^2 = (b - c)(b^2 + bc + c^2) \Rightarrow (b - c)(b^2 + bc + c^2 - a^2) = 0$.

Thus, we get: $b = c$, which indicates an isosceles triangle, or we get: $a^2 = c^2 + bc + b^2$, which does not indicate a triangle of any particular kind.

What then, can we say?

We can notice though, that the expression has $a^2 = c^2 + b^2$, which can be found in the cosine rule, where: $a^2 = b^2 + c^2 - 2bc \cos A$. So we may want to try the rule now.

How can we try the rule though?

We can put the rule this way, too: $2bc \cos A = b^2 + c^2 - a^2$.

And we can get: $a^2 = c^2 + bc + b^2 \Rightarrow bc = a^2 - c^2 - b^2 \Rightarrow -bc = b^2 + c^2 - a^2$.

So applying the rule now to the above, we can get: $-bc = 2bc \cos A \Rightarrow \cos A = -1/2$.

And we know: $\cos \pi/3 = 1/2$.   What then, about $\cos A = -1/2$?

We have a trig-identity where: $\cos (\pi - \theta) = -\cos \theta$, and have: $-\cos \pi/3 = -1/2$, too.   So?

So we get: **cos ($\pi - \pi/3$) = -cos $\pi/3$ = -1/2**.

Thus, we get: $A = \pi - \pi/3 = 2\pi/3$, which is $120°$.

So it can be any of the three cases below:

• It is an isosceles triangle with $b = c$.

• It is a triangle with $A = 2\pi/3$.

• It is an isosceles triangle with $A = 2\pi/3$.

Examples 3 in The Cosine Rule

0.   A road gets split into two branches making an angle of $5\pi/12$. And there are going to be two gas stations U and V, each of which will be located in each of the branch roads.

And as shown in the figure below, there are two towns X and Y, each of which is 30 Km away from the junction T where the branches begin. The angle XTY is $\pi/6$, and the angle YTV is $\pi/12$.

Find the minimum of the sum of the three distances of XU, UV, and VY, that is, the minimum of $XU + UV + VY$.

Fig. 0.0

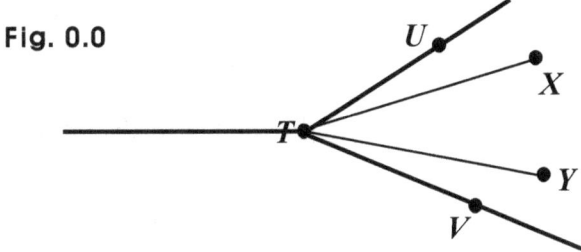

1.   Suppose again, a road gets split into two branches making an angle of $5\pi/12$. And there are going to be two gas stations U and V, each of which will be located in each of the branch roads. And a truck depot X is located between the branches, and is 30 Km away from the junction T where the branch roads begin.

Find the minimum distance of $(XU + UV + VX)$.

Suggestions or Solutions
To the **Problem** in the Example **0**

A road gets split into two branches making an angle of $5\pi/12$. And there are going to be two gas stations U and V, each of which will be located in each branch road. And as shown in the figure below, there are two towns X and Y, each of which is 30 Km away from the junction T where the branches begin. The angle XTY is $\pi/6$, and the angle YTV is $\pi/12$.

Find the minimum of the sum of the three distances, XU, UV, and VY, that is, the minimum of $XU + UV + VY$.

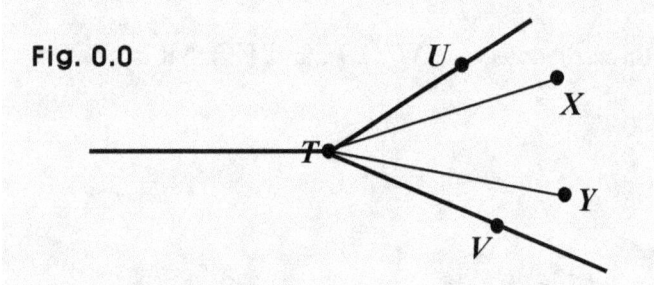

Fig. 0.0

We only know that the two gas stations will be somewhere in the branch roads, but we don't know yet where exactly the gas stations will be. So the exact positions of the two gas stations have not been determined yet. And in fact, we want to determine the positions so that the sum of the distances specified in the problem is the minimum.

Suppose to begin with, we want to go from Q to P, but have to visit a point in the line L below. What point then, do we have to visit in the line L if the total distance we travel is the minimum?

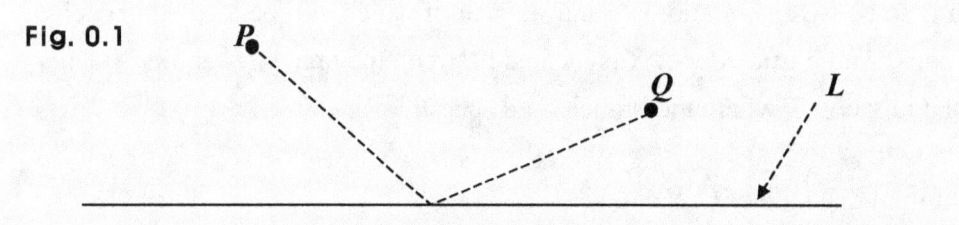

Fig. 0.1

Assuming the point is M, we can find M using the triangle inequality. How?

Fig. 0.2

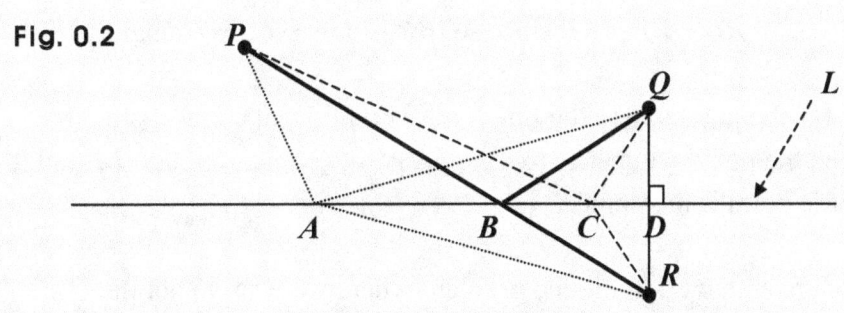

Suppose we put a point **R** as shown in the figure above, so that **QD = DR**, and **QR** is perpendicular to the line **L**.

Then, we get: **QB = RB**, **QA = RA**, and **QC = RC** because ΔQBR, ΔQAR, and ΔQCR are isosceles triangles.

So first, using the triangle inequality, we can get: **QA + AP > RP**. How come?

That's because: **QA + AP = RA + AP > RP** due to the triangle inequality.

And next, using the triangle inequality again, we can get: **QC + CP > RP**. How come?

That's because: **QC + CP = RC + CP > RP** due to the triangle inequality.

Now, we know **A** and **C** are arbitrary points in the line **L**.

So no matter what the two points **A** and **C** may be, we get:

QA + AP > RP, and **QC + CP > RP**. So what?

The length of **RP** is the minimum distance we can travel.

So the sum, **QB + BP** is the minimum, that is, **B** is the point **M**. How come?

That's because: $QB + BP = RB + BP = RP$ since $\triangle QBR$ is an isosceles triangle.

And thus, the point M is the point where the line L meets a line that connects the destination point P and the point symmetric to the start point Q about the line L. And in this case, the symmetric point is the point R.

So we can now use the fact above solving the problem where we want to get the minimum of the sum of the three distances, that is, the minimum of $XU + UV + VY$.

And thus, putting into the figure given, the angles given, together with some labels necessary, we can get:

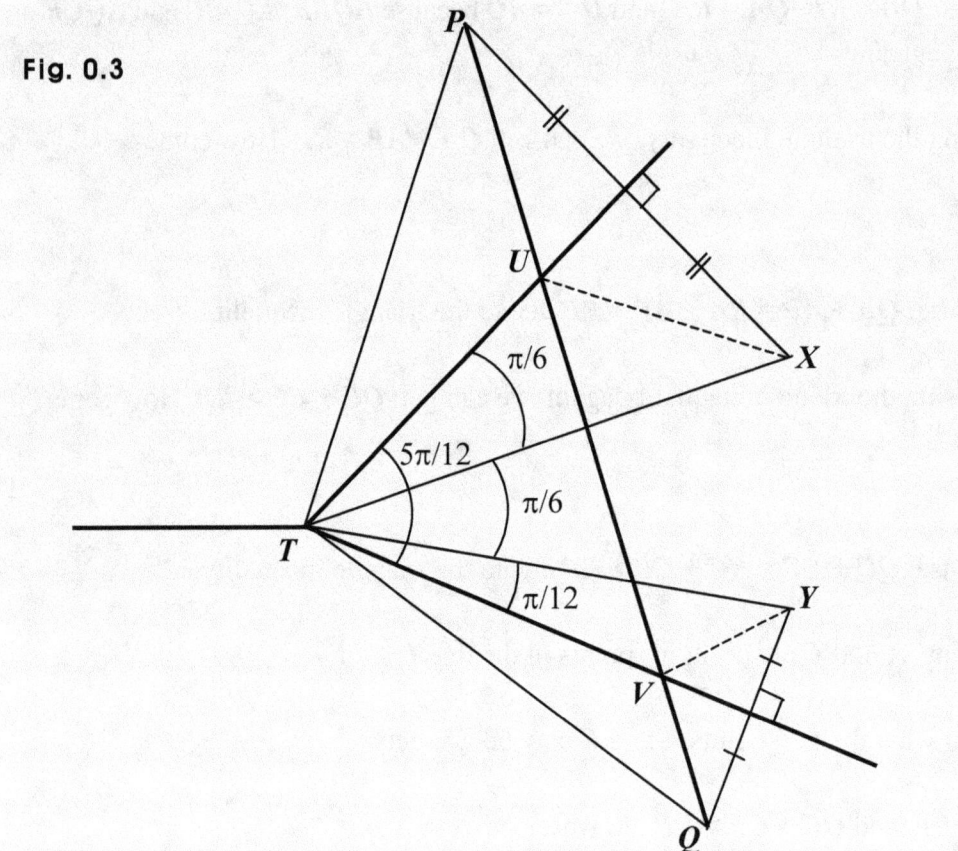

Fig. 0.3

So finding the length of PQ, we get the minimum of $XU + UV + VY$.

Now, to begin with, we have: $TX = TY = 30$. So we get: $PT = TQ = 30$, too.

That's because $\triangle PTX$ and $\triangle YTQ$ are isosceles triangles.

Next, the angle **PTU** is π/6, and the angle **QTV** is π/12.

That's also because **ΔPTX** and **ΔYTQ** are isosceles triangles.

So assuming A is the angle **PTQ**, we get: $A = 5\pi/12 + \pi/12 + \pi/6 = 8\pi/12 = 2\pi/3$.

How then, can we get the length of **PQ**?

We have the cosine rule, where: $a^2 = b^2 + c^2 - 2bc \cos A$, where A is the angle facing the side a, and is the angle between the two sides b and c.

So assuming $a = PQ$, $b = PT$, and $c = TQ$, we can say that $A = 2\pi/3$.

And we know: $\cos(\pi - \theta) = -\cos\theta$, and $\cos\pi/3 = 1/2$. So we can get:

$a^2 = b^2 + c^2 - 2bc \cos 2\pi/3 \Rightarrow a^2 = 30^2 + 30^2 - 2 \cdot 30 \cdot 30 \,(-\cos\pi/3) = 3 \cdot 30^2$.

And thus, the minimum is: $30\sqrt{3}$.

Suggestions or Solutions
To the **Problem** in the Example 1

Suppose again, a road gets split into two branches making an angle of 5π/12. And there are going to be two gas stations *U* and *V*, each of which will be located in each of the branch roads. And a truck depot *X* is located between the branches, and is 30 Km away from the junction *T* where the branch roads begin. Find the minimum distance of *XU* + *UV*+ *VX*.

Using the fact about the point *M* stated in the example 9, we can show the way below, the route that takes the minimum distance:

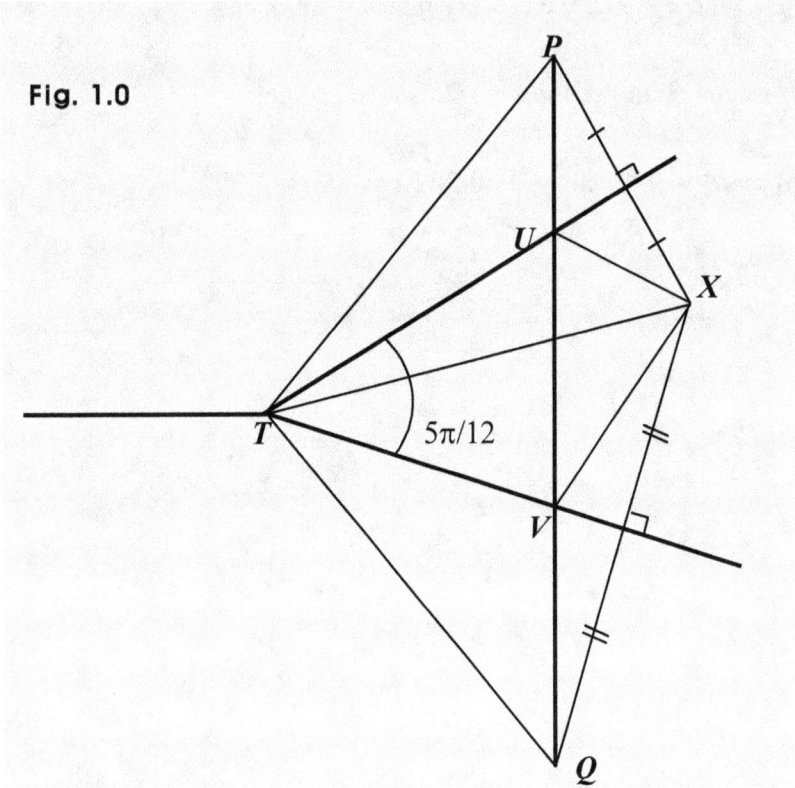

Fig. 1.0

So finding the length of *PQ*, we get the minimum of *XU* + *UV* + *VX*.

Now, to begin with, we have: *TX* = **30**.   So we get: *PT* = *TQ* = **30**, too.

That's because **ΔPTX** and **ΔQTX** are isosceles triangles.

Next, the angle **PTQ** is 10π/12, which is 5π/6.   How come?

That's also because **ΔPTX** and **ΔYTQ** are isosceles triangles.

The sum of the angle **UTX** and the angle **VTX** is the same as the sum of the angle **QTV** and the angle **PTU**.

So assuming A is the angle **PTQ**, we get: $A = 2 \cdot 5\pi/12 = 10\pi/12 = 5\pi/6$.

How then, can we get the length of **PQ**?

We have the cosine rule, where: $a^2 = b^2 + c^2 - 2bc \cos A$, where A is the angle facing the side a, and is the angle between the two sides b and c.

So assuming $a = PQ$, $b = PT$, and $c = TQ$, we can say that $A = 5\pi/6$.

And we know: $\cos(\pi - \theta) = -\cos\theta$, and $\cos \pi/6 = \frac{\sqrt{3}}{2}$.

So we can get:

$a^2 = b^2 + c^2 - 2bc \cos 5\pi/6 \Rightarrow a^2 = 30^2 + 30^2 - 2 \cdot 30 \cdot 30 (-\cos \pi/6) = 2 \cdot 30^2 (1 + \frac{\sqrt{3}}{2})$

$= 30^2 (2 + \sqrt{3})$.

And thus, the minimum is: $30\sqrt{2 + \sqrt{3}}$.

₇.**Areas of Triangles**

It seems quite easy to find the area of a triangle.   How do we find it though?

We can find it finding the area of a parallelogram, and then, taking half the area.

So the area of the parallelogram is of course, twice the area of the triangle we want to get. And finding the area of a parallelogram, we take the product of the length and the width or the height.

So let's now, for instance, find the area of a triangle below:

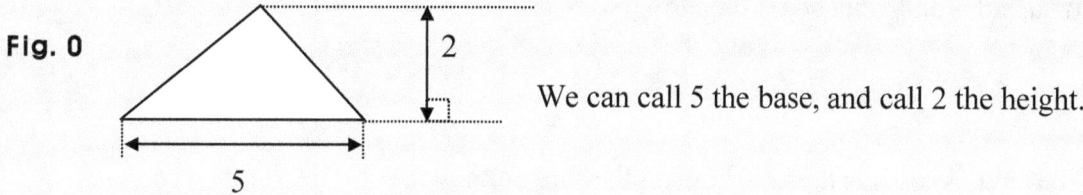

Fig. 0

We can call 5 the base, and call 2 the height.

Then first, we make a parallelogram twice as big as the triangle above. And making such a parallelogram, we add to the triangle the same triangle the way below:

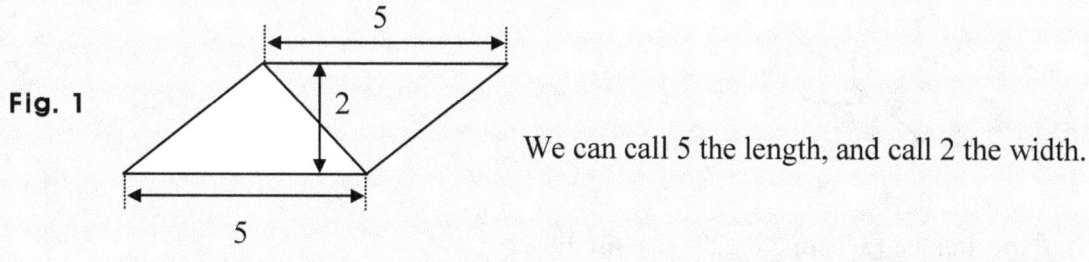

Fig. 1

We can call 5 the length, and call 2 the width.

And taking the area of the parallelogram above, we simply get: 5·2 = 10.   How come?

Cutting either triangle into two right triangles, and then, moving one the way below, we get a rectangle that has the same area the parallelogram has.

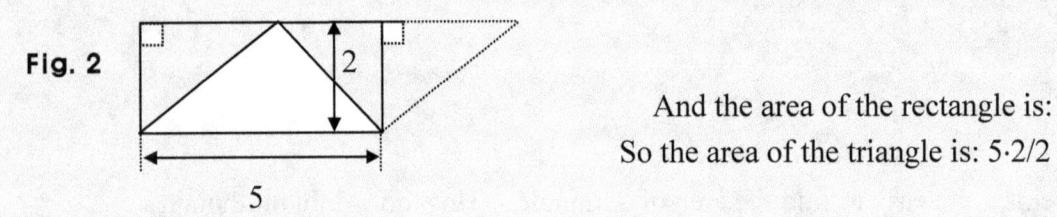

Fig. 2

And the area of the rectangle is: 5·2.
So the area of the triangle is: 5·2/2 = 5.

And thus, in short, taking the area of a triangle, we just take half the product of the base and the height.

What if though, we don't know the base or the height?

We can't get the area using the idea above.
If we know however, two sides and an angle, we can still get the area.

What sides and what angle though?

The angle known is the angle facing the side unknown.

That is to say that the angle is between the two sides known.

So for instance, we can get the area of the triangle below:

Fig. 3

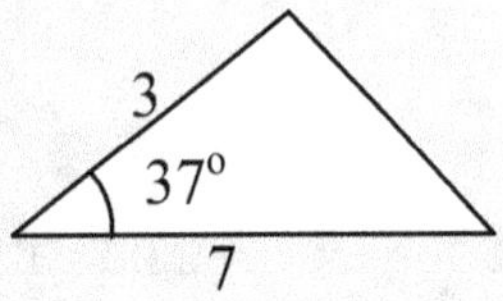

And the area is: **(3·7·sin 37°)/2 = (21·sin 37°)/2**.

And thus, assuming S is the area of the triangle below, we can get S the way below:

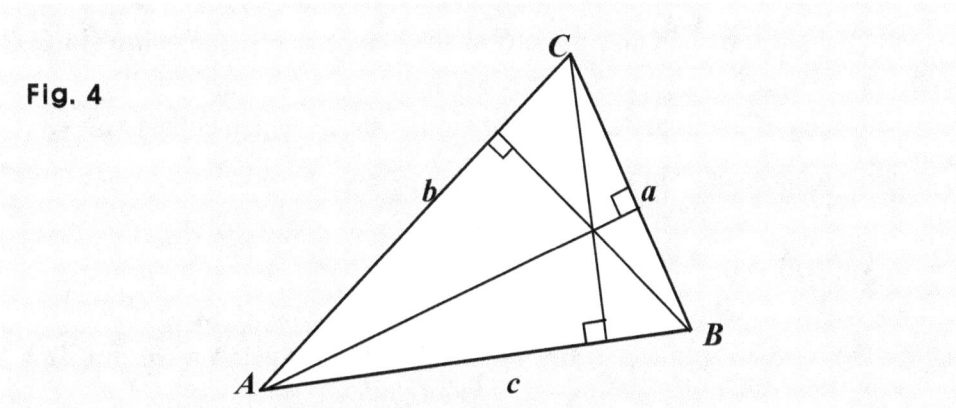

Fig. 4

Then, we get: $S = (ab \sin C)/2 = (bc \sin A)/2 = (ca \sin B)/2$. How come?

Assuming first, a is the base, we can take $b \sin C$ as the height, and get: $S = (ab \sin C)/2$.

Assuming next, b is the base, we can take $c \sin A$ as the height, and get: $S = (bc \sin A)/2$.

And assuming, c is the base, we can take $a \sin B$ as the height, and get: $S = (ca \sin B)/2$.

What if the triangle is an obtuse one as below?

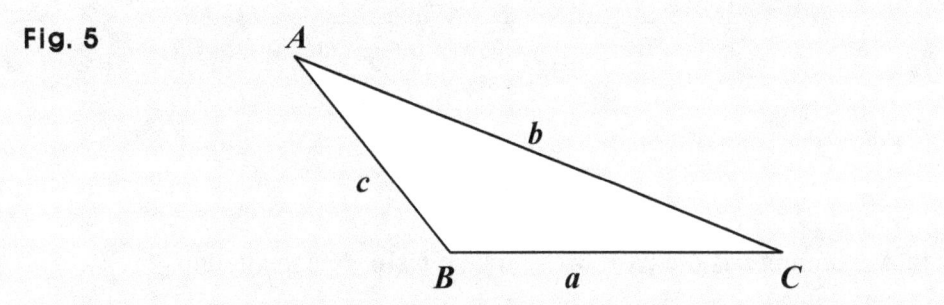

Fig. 5

We can put the triangle above the way below:

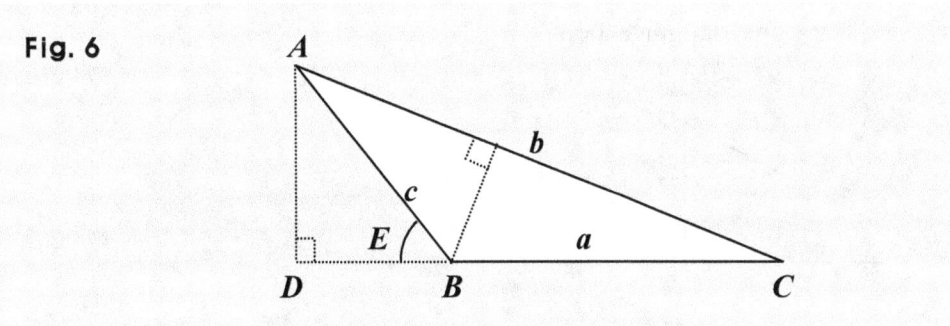

Fig. 6

Assuming first, **a** is the base, we can take **b sin C** as the height, and get: **S = (ab sin C)/2**.

Assuming again, **a** is the base, we can take as the height **c sin E**, which equals however, **c sin (π – B)**, which equals also, **c sin B**.

So we get: **S = (ac sin B)/2 = (ca sin B)/2**.

And assuming **b** is the base, we can take **c sin A** as the height, and get: **S = (bc sin A)/2**.

Can we not then, assume that **c** is the base?

Of course, we can. We can put it this way, too:

Fig. 7

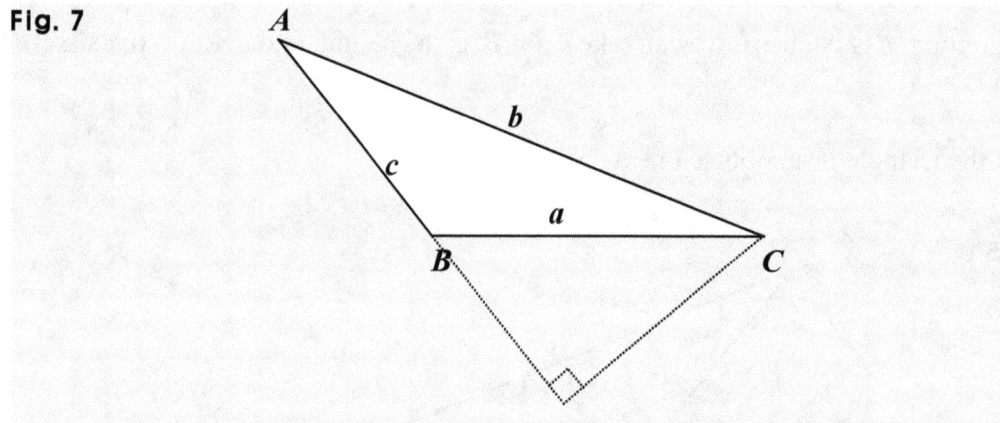

That is to say that assuming **c** is the base, we can take **b sin A** as the height.

Then, we get: **S = (bc sin A)/2**.

What if the triangle is a right triangle as below?

Fig. 8

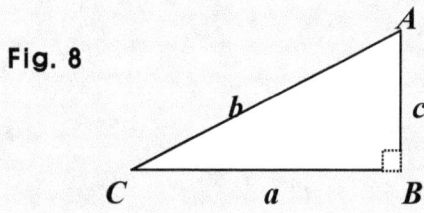

Assuming first, *a* is the base, we can take *c* **sin** *B* as the height. How come?

We know *B* is π/2, which is 90°. So we get: **sin** *B* = **1**. And thus, we get: *S* = (*ac* **sin** *B*)/2.

And we can put it the way below, too:

Assuming *c* is the base, we can take *a* **sin** *B* as the height since we have: **sin** *B* = **1**.

And thus, we get: *S* = (*ac* **sin** *B*)/2. *a* is the base, we can take *c* **sin** *B* as the height.

And assuming again, *a* is the base, we can take *b* **sin** *C* as the height, which is *c*, and thus, we get: *S* = (*ab* **sin** *C*)/2. What then, about using **sin** *A*?

Assuming *c* is the base, we can take *b* **sin** *A* as the height, which is *a*, and therefore, we get: *S* = (*bc* **sin** *A*)/2.

So in short, given the angle between two sides known, we can use: *S* = (*bc* **sin** *A*)/2, where *A* is the angle between *b* and *c*, of course.

What if we are given three sides only? That is, three sides with no angle are given.

We can get the area using the three sides only.

That's because three line segments makes one triangle only. So given the three line segments, we should be able to find the area of the triangle. Some one did in fact, find it.

And it is called Heron's formula, named after Heron of Alexandria.

Assuming *a*, *b*, and *c* are three sides in a triangle, and *S* is the area of it, we get:

$$S = \sqrt{t(t-a)(t-b)(t-c)} \text{ where } 2t = a + b + c, \text{ that is, } t = (a + b + c)/2.$$

And another formula was found by a Chinese mathematician, too, and is as follows:

$S = \frac{1}{2}\sqrt{a^2c^2 - (\frac{a^2+c^2-b^2}{2})^2}$, which is thus, equivalent to the one above, but has been found independently from the one above.

And what's more important to us is of course, how the formula can be made.

So let's see now, how it is the case.

It is in fact, from the formula where $S = (bc \sin A)/2$.

So using the formula above, we put S in terms of the three sides a, b, and c only.

We don't get it though, using the formula only, of course.

Getting the formula, we use a tool called a trig-identity and another tool called the cosine rule, which is: $a^2 = b^2 + c^2 - 2bc \cos A$.

And the trig-identity is: $\sin^2 A + \cos^2 A = 1$ for any angle A.

And of course, $S = (bc \sin A)/2$ is a tool in math, too.

So using all the three tools above, we can put S in terms of the three sides a, b, and c.

Now, we have:

$S = (bc \sin A)/2, \quad \sin^2 A + \cos^2 A = 1, \quad$ and $a^2 = b^2 + c^2 - 2bc \cos A$.

And thus, kneading the mixture of the three above, we get to come up with the one called Heron's formula. So what we are going to do is algebra.

After all, we get to put somehow $\sin A$ in terms of a, b, and c only.

To begin with, we can see that the third tool has the three sides.

How then, can we put $\sin A$ in terms of a, b, and c only?

There is a bridge between $\sin A$ and the three sides. What then, is the bridge?

It is the tool where $\sin^2 A + \cos^2 A = 1$.

So what we do is that we solve first, for $\cos A$ the equation $a^2 = b^2 + c^2 - 2bc \cos A$, and then, put it into the bridge above, isn't it?

We do solve it for $\cos A$, but it's not a good idea to it directly into the bridge, because the subsequent calculations will be quite messy. How then, can we avoid such a mess?

We have lots of tools to work with in math.
Among those, we have a tool called a factorization, called factoring, too.

And the tool is this: $x^2 - y^2 = (x + y)(x - y)$.

So to begin with, we get: $a^2 = b^2 + c^2 - 2bc \cos A \Rightarrow \cos A = \frac{b^2+c^2-a^2}{2bc}$.

And next, we get: $\sin^2 A + \cos^2 A = 1 \Rightarrow \sin^2 A = 1 - \cos^2 A = (1 + \cos A)(1 - \cos A)$.

And we have this, too: $\cos A = \frac{b^2+c^2-a^2}{2bc}$. So next, we get:

$$\sin^2 A = (1 + \cos A)(1 - \cos A) = (1+\frac{b^2+c^2-a^2}{2bc})(1-\frac{b^2+c^2-a^2}{2bc}) = (\frac{2bc+b^2+c^2-a^2}{2bc})(\frac{2bc-b^2-c^2+a^2}{2bc}),$$

which is not quite simple. We can notice though, that:

$$2bc + b^2 + c^2 = (b + c)^2, \text{ and } 2bc - b^2 - c^2 = -(b^2 - 2bc + c^2) = -(b - c)^2.$$

So we get: $\sin^2 A = (\frac{(b+c)^2-a^2}{2bc})(\frac{a^2-(b-c)^2}{2bc}) = \frac{1}{4b^2c^2} \{(b + c)^2 - a^2\}\{a^2 - (b - c)^2\}$.

Meanwhile:

$$(b + c)^2 - a^2 = (b + c + a)(b + c - a), \text{ and } a^2 - (b - c)^2 = (a + b - c)(a - b + c).$$

So we get: $\sin^2 A = \frac{1}{4b^2c^2}(a+b+c)(b+c-a)(a-b+c)(a+b-c)$.   Now, what?

In the formula we are proving now, we have: $2t = a + b + c$.

So it is time to do substitutions. And doing substitutions, we get:

$b+c-a = 2t-2a$, $a-b+c = 2t-2b$, and $a+b-c = 2t-2c$.

So we get: $(a+b+c)(b+c-a)(a-b+c)(a+b-c) = 2t(2t-2a)(2t-2b)(2t-2c)$
$= 2^4 t(t-a)(t-b)(t-c)$.

Thus, we get: $\sin^2 A = \frac{2^4}{4b^2c^2}t(t-a)(t-b)(t-c) = \frac{2^2}{b^2c^2}t(t-a)(t-b)(t-c)$.

So we get: $\sin A = \frac{2}{bc}\sqrt{t(t-a)(t-b)(t-c)}$, because $0 < A < \pi \Rightarrow \sin A > 0$, since the angle A is an angle in a triangle.

Now, we have: $S = (bc \sin A)/2 = \frac{bc}{2} \sin A$.

So we get: $S = \sqrt{t(t-a)(t-b)(t-c)}$ where $t = \frac{1}{2}(a+b+c)$.

Examples in Areas of Triangles

Doing the examples $0 - 2$, refer to the triangle ABC below:

Fig. 0

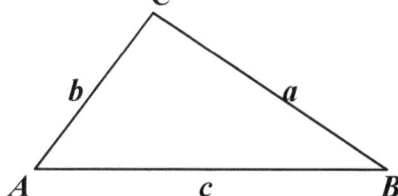

0.   Find the area of a triangle ABC where $A = \pi/3$, $C = \pi/4$, and $c = 10$.

1.   Suppose in a triangle ABC where $a = 2\sqrt{17}$, $b = 8$, and $c = 10$, a point U is in the side c, another point V is in the side b, the length of the line segment AU is x, and the length of the line segment AV is y.

1.0.   What then, is the value of the product xy if the area of the triangle AUV is half the area of the triangle ABC?

1.1.   What is the minimum length of the line segment UV if the area of AUV is half the area of ABC?

2.   Suppose S is the area of a triangle **ABC**, **R** is the radius of its circumcircle, and **r** is the radius of its inscribing circle.

2.0.   Show that: $r = \frac{2S}{a+b+c}$, $4SR = abc$, and $S = 2R^2 (\sin A)(\sin B)(\sin C)$.

2.1.   Find **r** and **R** assuming $a = 13$, $b = 14$, and $c = 15$.

3.   Find the area of a tetragon **ABCD** below assuming $B = \pi/3$, and $C = 5\pi/12$.

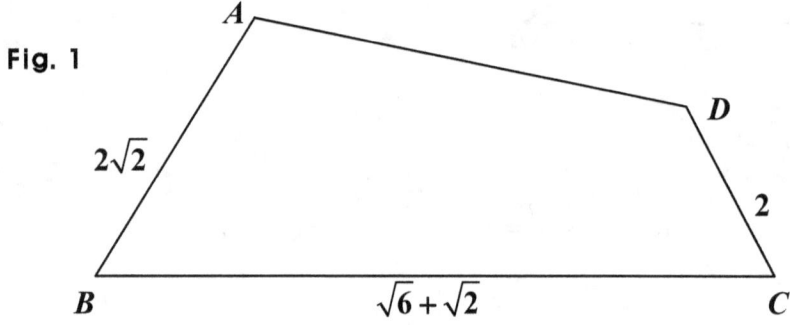

Fig. 1

Suggestions or Solutions

To the **Problem** in the Example **0**

Find the area of a triangle ABC where $A = \pi/3$, $C = \pi/4$, and $c = 10$.

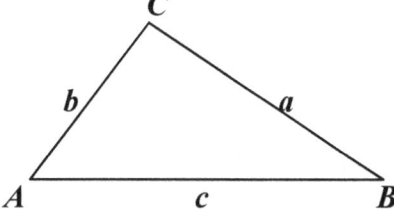

Fig. 0.0

Assuming d is the distance from the point B to b, we get: $d = 10 \sin 60° = 5\sqrt{3}$.

So assuming s is the area, and b is the base, we get: $s = \frac{5b\sqrt{3}}{2}$.

And we can put b this way: $10 \cos 60° + a \cdot \cos 45°$. And getting a, we get:

$$\frac{a}{\sin A} = \frac{c}{\sin C} \Rightarrow a = \frac{c}{\sin C} \cdot \sin A = \frac{10}{\sqrt{2}/2} \cdot \frac{\sqrt{3}}{2} = \frac{20}{\sqrt{2}} \cdot \frac{\sqrt{3}}{2} = \frac{10\sqrt{3}}{\sqrt{2}} = \frac{10\sqrt{6}}{2} = 5\sqrt{6}.$$

So we get: $b = 10 \cos 60° + a \cdot \cos 45°$

$$= 10 \cdot \tfrac{1}{2} + a \tfrac{\sqrt{2}}{2} = \frac{10 + a\sqrt{2}}{2} = \frac{10 + 5\sqrt{6}\sqrt{2}}{2} = \frac{10 + 10\sqrt{3}}{2} = 5(1 + \sqrt{3}).$$

And thus, we get: $s = \frac{5b\sqrt{3}}{2} = \frac{25(1+\sqrt{3})\sqrt{3}}{2} = \frac{25(3+\sqrt{3})}{2}$.

If not quite sure of the idea behind the processes above, follow the steps below:

To begin with, using the information given, we can put the triangle the way below:

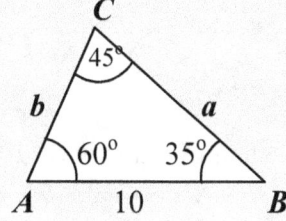

Fig. 0.1

Basically, the area of a triangle is: the base times the height over two.
And we want to determine the base first.

So for instance, in this example, finding the side b, and the height, we can get the area.
What is the height though?

We can get the distance from the point at B to the side b.

Then, the distance is the height since b has been chosen to be the base.

And the distance is in fact: $10 \sin 60^\circ$, which is $5\sqrt{3}$ since $\sin 60^\circ = \frac{\sqrt{3}}{2}$.

So assuming b is the base, we can get: $\frac{1}{2} b \cdot 5\sqrt{3}$, which is: $\frac{5b\sqrt{3}}{2}$, which is the area.
How then, can we get the side b?

We have a tool called the sine rule, where: $\dfrac{a}{\sin A} = \dfrac{b}{\sin B} = \dfrac{c}{\sin C} = 2R$, where R is the

radius of the circumcircle to the triangle ABC.

So the circle passes through all the three vertices of the triangle ABC.
And we know the three angles, and the value of c. So?

So using $\dfrac{b}{\sin B} = \dfrac{c}{\sin C}$, we can get b.

The angle B is however 35°, the sine of which cannot be readily found unless we use a
calculator. What else then, can we do to get the side b?

We can readily get $\cos 60^\circ$ and $\cos 45^\circ$. So we can put b this way: $10 \cos 60^\circ + a \cdot \cos 45^\circ$.
How come?

Fig. 0.2

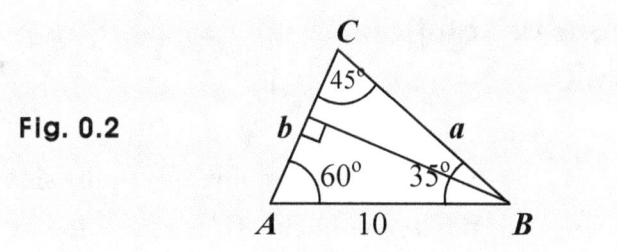

So we can set: $b = 10 \cos 60° + a \cdot \cos 45°$.   What then, about a?

We know $\sin 60° = \frac{\sqrt{3}}{2}$, and $\sin 45° = \frac{\sqrt{2}}{2}$.   So?

So we can use the sine rule to get a.   And thus, using the rule, we get:

$$\frac{a}{\sin A} = \frac{c}{\sin C} \Rightarrow a = \frac{c}{\sin C} \cdot \sin A = \frac{10}{\sqrt{2}/2} \cdot \frac{\sqrt{3}}{2} = \frac{20}{\sqrt{2}} \cdot \frac{\sqrt{3}}{2}$$

$$= \frac{10\sqrt{3}}{\sqrt{2}} = \frac{10\sqrt{6}}{2} = 5\sqrt{6}.$$ So we get: $a = 5\sqrt{6}$.

So putting it into this: $b = 10 \cos 60° + a \cdot \cos 45°$, where $\cos 60° = 1/2$, and $\cos 45° = \frac{\sqrt{2}}{2}$,

we get: $b = \frac{10}{2} + \frac{a\sqrt{2}}{2} = \frac{10 + a\sqrt{2}}{2} = \frac{10 + 5\sqrt{6}\sqrt{2}}{2} = \frac{10 + 10\sqrt{3}}{2} = 5(1 + \sqrt{3})$.

And we know that the area is: $\frac{5b\sqrt{3}}{2}$.

So assuming s is the area, we get: $s = \frac{5b\sqrt{3}}{2} = \frac{25(1+\sqrt{3})\sqrt{3}}{2} = \frac{25(3+\sqrt{3})}{2}$.

Suggestions or Solutions
To the **Problem 0** in the Example 1

Suppose in a triangle ABC where $a = 2\sqrt{17}$, $b = 8$, and $c = 10$, a point U is in the side c, another point V is in the side b, the length of the line segment AU is x, and the length of the line segment AV is y.

What then, is the value of the product xy if the area of the triangle AUV is half the area of the triangle ABC?

Fig. 1.0.0

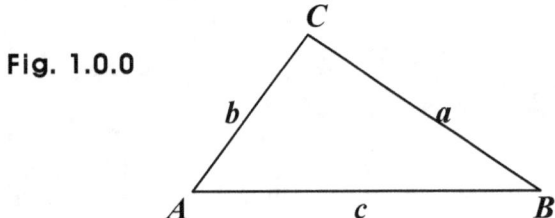

To begin with, using the information given, we can put the triangle the way below:

Fig. 1.0.1

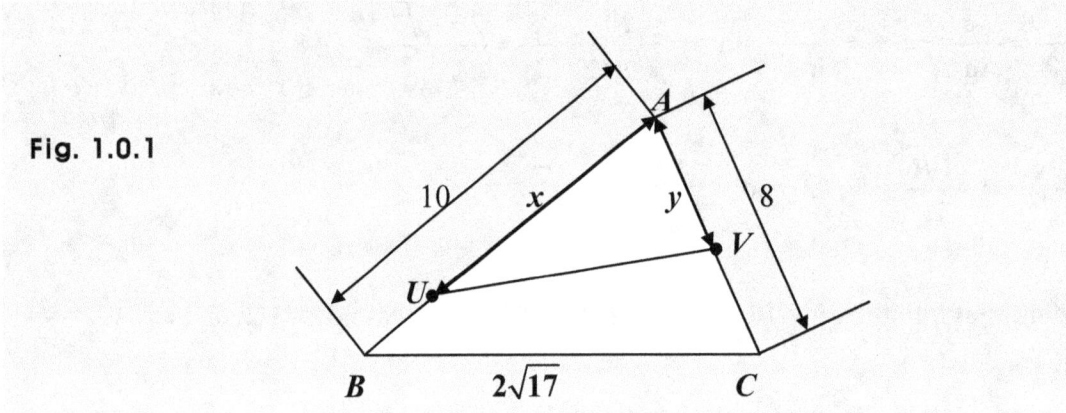

Assuming T is the area of $\triangle AUV$, and t is the area of $\triangle ABC$, we want determine the value of the product xy so that we get: $T = 2t$. What then, about the value of xy?

In fact, finding the area of $\triangle AUV$, we get to see the value of xy.

That's because the area will have the value of xy. So let's now, find the area of $\triangle AUV$.

Assuming *x* is the base, we can put the area of $\triangle AUV$ this way: $t = (xy \sin A)/2$.

And assuming the side *b* is the base, we can put the area of $\triangle ABC$ the way below:

$T = (bc \sin A)/2$ where $b = 8$, and $c = 10$.

So we get: $T = 40 \sin A$.   How then, can we get the **sin** *A*?

We don't really need it. Why not?

What we want is *xy*.   So we can use this instead: $T = 2t$.

Thus, using it, we get:

$T = 2t \Rightarrow 40 \sin A = 2(xy \sin A)/2 = xy \sin A \Rightarrow 40 \sin A = xy \sin A \Rightarrow xy = 40$.

Suggestions or Solutions
To the **Problem 1** in the Example 1

Suppose in a triangle ABC where $a = 2\sqrt{17}$, $b = 8$, and $c = 10$, a point U is in the side c, another point V is in the side b, the length of the line segment AU is x, and the length of the line segment AV is y.

What then, is the minimum length of the line segment UV if the area of AUV is half the area of ABC?

Fig. 1.1.0

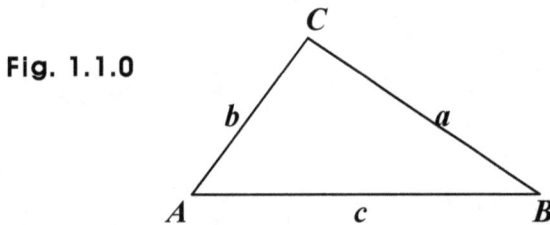

Setting: $s = UV$, we can get:: $s^2 = x^2 + y^2 - 2xy \cos A$.

And we get: $\cos A = (b^2 + c^2 - a^2)/(2bc) = (64 + 100 - 68)/160 = 96/160 = 3/5$.

And we have: $xy = 40$. So we can set: $y = 40/x$.

Thus, we get: $s^2 = x^2 + y^2 - 2xy \cos A = x^2 + (40/x)^2 - 2 \cdot 40(3/5) = x^2 + 40^2/x^2 - 48$.

So assuming $u = x^2$, and $v = s^2$, we can set: $v = u + 40^2/u - 48$ for $u > 0$.

And assuming, $h(u) = u + 40^2/u$, $f(u) = u$, and $g(u) = 40^2/u$, we can set:

$v = h(u) = f(u) + g(u)$ for $u > 0$.

And referring to the graph above, we can see that when $u = 40^2/u$, h gets its minimum value. And we get: $u = 40^2/u \Rightarrow u^2 = 40^2 \Rightarrow u = 40$.

And thus, when $u = 40$, v gets its minimum, which is $40 + 40^2/40 - 48 = 80 - 48 = 32$.

And we know: $v = s^2$, which is the square of the length of UV.

So the minimum length of UV is: $\sqrt{32} = 4\sqrt{2}$.

If not quite sure of the idea behind the processes above, follow the steps below:

To begin with, using the information given, we can put the triangle the way below:

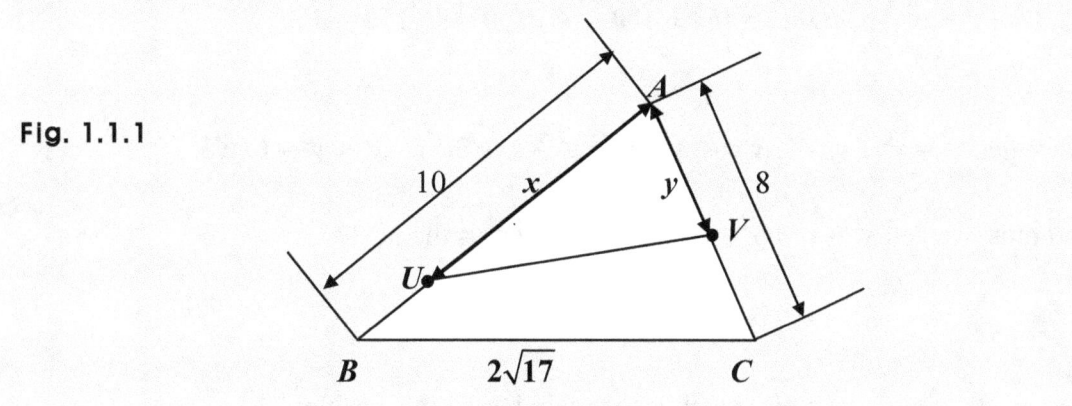

Fig. 1.1.1

It can be the case, of course, *x* and *y* can change keeping *xy* constant to be 40.

That is to say that *U* and *V* can move along the sides respectively so that the length of the line segment *UV* can change keeping the area of $\triangle AUV$ half the area of $\triangle ABC$, that is, keeping unchanged the equation *T = 2t*.

So we may want to keep track of the values of the line segment *UV* as other values vary.

And thus, we may want to take the length of *UV* as a variable.

So for instance, we can set: *s = UV*.
And we know that *x* and *y* vary keeping *xy* 40 though, of course.

How then, can we relate *s*, *x*, and *y*?

We have the cosine rule, where: $p^2 = q^2 + r^2 - 2qr \cos P$ for $\triangle PQR$.

So we can set: $s^2 = x^2 + y^2 - 2xy \cos A$ for $\triangle AUV$.   How then, can we get **cos A**?

We have this, too: $a^2 = b^2 + c^2 - 2bc \cos A$, where $a = 2\sqrt{17}, b = 8$, and $c = 10$.

So we can get $\cos A$ this way:

$\cos A = (b^2 + c^2 - a^2)/(2bc) = (64 + 100 - 68)/160 = 96/160 = 3/5$.

So we get: $s^2 = x^2 + y^2 - 2xy \cos A = x^2 + y^2 - 2xy(3/5) = x^2 + y^2 - 6xy/5$.

And thus, we get: $s^2 = x^2 + y^2 - 6xy/5$. What then, is the next?

We have this, also: $xy = 40$, which is the main key to this problem.

So we can get: $y = 40/x$.

Then, we get: $s^2 = x^2 + y^2 - 6xy/5 = x^2 + 40^2/x^2 - (6/5)x40/x = x^2 + 40^2/x^2 - 48$.

That is, we get: $s^2 = x^2 + 40^2/x^2 - 48$.   And we know x^2 is always positive.

So assuming $u = x^2$, and $v = s^2$, we can set: $v = u + 40^2/u - 48$ for $u > 0$.

And thus, getting the minimum of v, we can get the minimum of s, which is the minimum of UV, of course.   How then, can we get the minimum of v?

We can notice that even if u changes 48 remains 48, and thus, remains constant.

So we can keep track of $u + 40^2/u$ to find the minimum of v.

And assuming, $h(u) = u + 40^2/u, f(u) = u$, and $g(u) = 40^2/u$, we can set:

$v = h(u) = f(u) + g(u)$ for $u > 0$.   And putting it in a graph, we get:

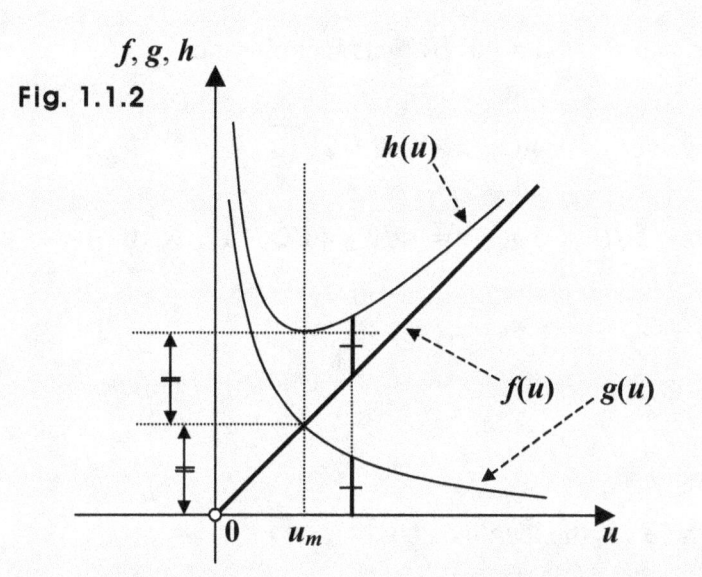

f, g, h

Fig. 1.1.2

h(u)

f(u)   g(u)

0   u_m u

So we can see that when $f(u) = g(u)$, that is, $u = 40^2/u$, h gets its minimum value.

That is, if $u = u_m$, h gets its minimum.

And thus, finding u_m in the case above, we get: $u_m = 40^2/u_m \Rightarrow u_m^2 = 40^2 \Rightarrow u_m = 40$.

So when $u = 40$, h gets its minimum.

And we know: $v = u + 40^2/u - 48$, and $h(u) = u + 40^2/u$.

So we can set: $v = h(u) - 48$, where 48 is not subject to u.

And thus, when $u = 40$, v gets its minimum, which is $40 + 40^2/40 - 48 = 80 - 48 = 32$.

That is, the minimum of v is 32.

And we know: $v = s^2$, which is the square of the length of UV.

So the minimum length of UV is $\sqrt{32} = 4\sqrt{2}$.

By the way, what are x and y when UV gets its minimum?

We know that we set: $u = x^2$, and that when $u = 40$, *UV* gets its minimum.

So when *UV* gets its minimum, we get: $x^2 = 40 \Rightarrow x = \sqrt{40} = 2\sqrt{10}$.

And we have: $xy = 40$. So when $x = 2\sqrt{10}$, we get: $y = 40/x = 40/2\sqrt{10} = 2\sqrt{10}$.

And we can get the same the way below, too:

We have: $s^2 = x^2 + y^2 - 6xy/5$, where *s* is the length of *UV*.

And we can get: $s^2 = x^2 + y^2 - 6xy/5 = x^2 + y^2 - 2xy + 2xy - 6xy/5 = (x - y)^2 + 4xy/5$.

And we know that $xy = 40$. So we get : $s^2 = (x - y)^2 + 32$.

And thus, when $x = y$, s^2 gets its minimum, which is 32.

And we know s^2 is the square of the length of *UV*.

So when $x = y$, the length of *UV* gets its minimum, which is: $\sqrt{32} = 4\sqrt{2}$.

Suggestions or Solutions
To the **Problem 0** in the Example **2**

Assuming S is the area of a triangle ABC, R is the radius of its circumcircle, and r is the radius of its inscribing circle, show that:

$$r = \frac{2S}{a+b+c}, \ 4SR = abc, \text{ and } S = 2R^2 (\sin A)(\sin B)(\sin C).$$

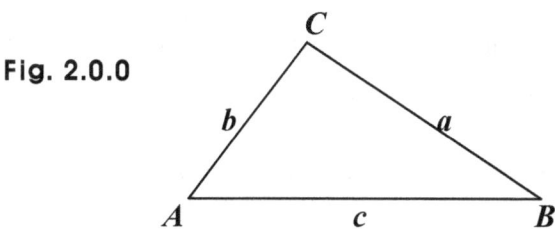

Fig. 2.0.0

To begin with, using the information given, we can put the triangle the way below:

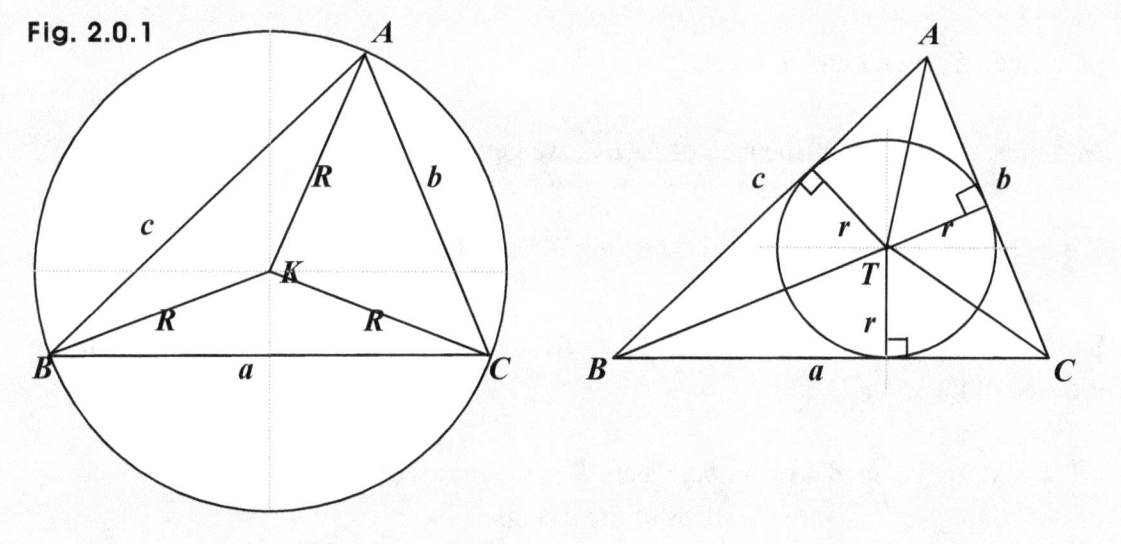

Fig. 2.0.1

Then first, what do we mean by $r = \frac{2S}{a+b+c}$?

We can put it this way, too: $S = \frac{(a+b+c)r}{2}$, where r is the radius of the circle inscribing to the triangle ABC. What do we mean by though, the circle inscribing?

It is the circle to which the three sides of the triangle *ABC* are tangent.

So we can say that the inscribing circle is tangent to all the three sides of the triangle.

And if a circle is tangent to all the sides of a polygon as a square, the circle can be said to inscribe the polygon. Not every polygon has its inscribing circle though. A rectangle cannot have it. Every triangle though, has its inscribing circle.

Now, the equation above is in fact, quite straightforward because *r* is the height of each of the three triangles, ΔTAB, ΔTBC, and ΔTCA in the figure above.

Assuming for instance, *a* is the base of ΔTBC, and S_a is the area of ΔTBC, we can set:

$S_a = ar/2$.

And the same is true, too, for the other two triangles, ΔTAB, and ΔTCA.

So we get: $S_b = br/2$ and $S_c = cr/2$.

And thus, assuming *S* is the area of ΔABC, we get:

$$S = S_a + S_b + S_c = ar/2 + br/2 + cr/2 = (a + b + c)r/2 \Rightarrow S = \frac{(a+b+c)r}{2} \Rightarrow r = \frac{2S}{a+b+c}.$$

So knowing the radius of the inscribing circle, along with all the three sides, we can get the area of the triangle.

Let's next, move on to **4SR = abc**, where **R** is the radius of the circumcircle, which passes through all the three vertices of the triangle.

And if a circle passes through all the vertices of a polygon as a pentagon, the circle can be said to circumscribe the polygon. Not every polygon however, has its circumscribing circle, called circumcircle, for short. For instance, some quadrangles (tetragons) and some pentagons cannot have it.

Now, what do we mean by **4SR = abc**?

We can put it this way, too: $S = \frac{abc}{4R}$, which is the area of the triangle ABC.

And of course, we can put the area in another way, too.

Knowing two sides and the angle between the two sides, we can get the area.

For instance, if we know b and c, together with the angle A, we can set: $S = (bc \sin A)/2$.

What then, about R, which is the radius of the circumcircle?

What equation has such a radius R, the radius of the circumcircle?

We have the sine rule, where: $\dfrac{a}{\sin A} = \dfrac{b}{\sin B} = \dfrac{c}{\sin C} = 2R$.

So using the sine rule, we can get: $\dfrac{a}{\sin A} = 2R \Rightarrow \sin A = \dfrac{a}{2R}$.

Thus, we get: $S = (bc \sin A)/2 = \dfrac{abc}{4R} \Rightarrow 4SR = abc$.

So knowing the radius of the circumcircle, along with all the three sides, we can get the area of the triangle, too.

And let's next, move on to the last case where $S = 2R^2 (\sin A)(\sin B)(\sin C)$.

We know S is the area of the triangle ABC.

So the equation where $S = 2R^2 (\sin A)(\sin B)(\sin C)$ is just another expression for the area of the triangle ABC.
And the area is put in terms of the radius of the circumcircle and the all the three angles of the triangle ABC.

How then, can we get the equation?

The equation is in terms of the sines of all the three angles.
So we can think of the sine rule. And we have: **$4SR = abc$**.

So first, putting all the sides in terms of the sines, we get:

$a = 2R \sin A$, $b = 2R \sin B$, and $c = 2R \sin C$.

So next, putting all the above into the equation where **$4SR = abc$**, we get:

$4SR = abc = (2R \sin A)(2R \sin B)(2R \sin C) = 8R^3 (\sin A)(\sin B)(\sin C)$

$\Rightarrow 4SR = 8R^3 (\sin A)(\sin B)(\sin C) \Rightarrow S = 2R^2 (\sin A)(\sin B)(\sin C)$.

And we can get the same the way below, too:

We can put **S** this way, too: **$S = (bc \sin A)/2$**.

And using the sine rule again, we get: **$b = 2R \sin B$ and $c = 2R \sin C$.**

So we get: **$S = (bc \sin A)/2 = (2R \sin B)(2R \sin C)(\sin A)/2 = 2R^2 (\sin A)(\sin B)(\sin C)$.**

In short:

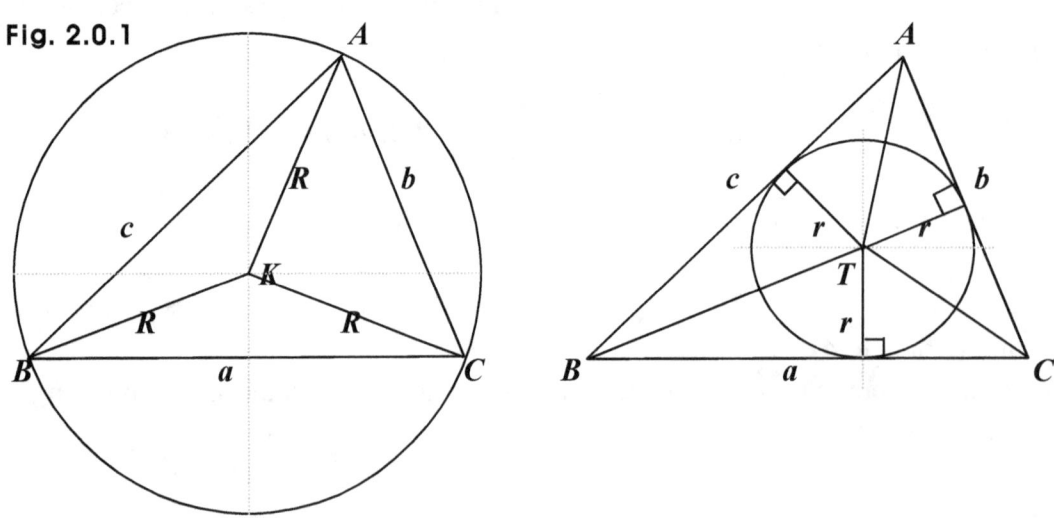

Fig. 2.0.1

Referring to the figure above, we can get:

$$S = ar/2 + br/2 + cr/2 = (a + b + c)r/2 \Rightarrow S = \frac{(a+b+c)r}{2} \Rightarrow r = \frac{2S}{a+b+c}.$$

Next, we can set: $S = (bc \sin A)/2$. And using the sine rule, we can get:

$$\frac{a}{\sin A} = 2R \Rightarrow \sin A = \frac{a}{2R}. \quad \text{So we get } S = (bc \sin A)/2 = \frac{abc}{4R} \Rightarrow 4SR = abc.$$

And next, using the sine rule again, we get: $a = 2R \sin A$, $b = 2R \sin B$, and $c = 2R \sin C$.

So next, putting all the above into the equation where $4SR = abc$, we get:

$$4SR = abc = (2R \sin A)(2R \sin B)(2R \sin C) = 8R^3 (\sin A)(\sin B)(\sin C)$$
$$\Rightarrow 4SR = 8R^3 (\sin A)(\sin B)(\sin C) \Rightarrow S = 2R^2 (\sin A)(\sin B)(\sin C).$$

And we can put S this way, too: $S = (bc \sin A)/2$.

And using the sine rule again, we get: $b = 2R \sin B$ and $c = 2R \sin C$.

So we get: $S = (bc \sin A)/2 = (2R \sin B)(2R \sin C)(\sin A)/2 = 2R^2 (\sin A)(\sin B)(\sin C)$.

Suggestions or Solutions
To the **Problem 1** in the Example **2**

Suppose again, S is the area of a triangle ABC, R is the radius of its circumcircle, and r is the radius of its inscribing circle.
What then, are r and R if $a = 13$, $b = 14$, and $c = 15$?

Fig. 2.1.0

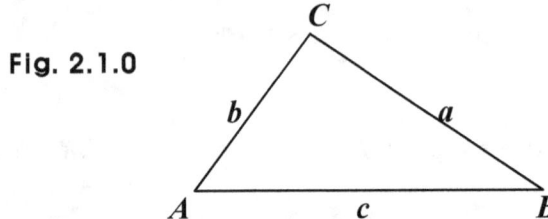

We have: $S = \sqrt{t(t-a)(t-b)(t-c)}$ where $2t = a + b + c$.

So first, we get: $2t = a + b + c = 13 + 14 + 15 = 42 \Rightarrow t = 21$.   Thus next, we can get:

$t(t-a)(t-b)(t-b) = 21(21-13)(21-14)(21-15) = 21 \cdot 8 \cdot 7 \cdot 6 = 2^4 \cdot 3^2 \cdot 7^2 = 84^2$.

So we get: $S = 84$.   And we have: $r = \frac{2S}{a+b+c}$ and $4SR = abc$. So we get:

$r = 2 \cdot 84/42 = 4$, and $R = abc/4S = 13 \cdot 14 \cdot 15/(4 \cdot 84) = 13 \cdot 2 \cdot 7 \cdot 3 \cdot 5/(2 \cdot 2 \cdot 7 \cdot 3 \cdot 4) = 65/8$.

If not quite sure of the idea behind the processes above, follow the steps below:

To begin with, using the information given, we can put the triangle the way below:

Fig. 2.1.1

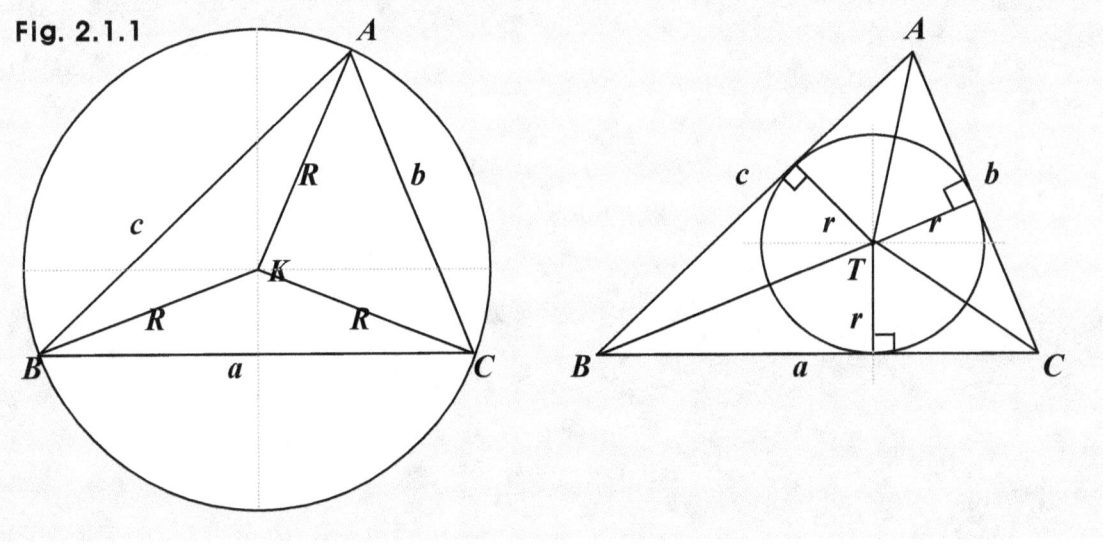

We have: $r = \frac{2S}{a+b+c}$ and $4SR = abc$, proved in the previous example.

So finding S, we can get r and R.   How then, can we get S, i.e., the area of the triangle?

None of the angles are given. So we cannot use the formula $S = (bc \sin A)/2$.
We can find though, the angles using the cosine rule as $a^2 = b^2 + c^2 - 2bc \cos A$.

Let's this time however, get the area without using any of the angles.

We have a tool called Heron's formula, which says:

Assuming $2t = a + b + c$, we get: $S = \sqrt{t(t-a)(t-b)(t-c)}$ for $\triangle ABC$.

So getting t first, we get: $2t = a + b + c = 13 + 14 + 15 = 42 \Rightarrow t = 21$.

Thus next, we can get : $t(t-a)(t-b)(t-b) = 21(21-13)(21-14)(21-15)$
$= 21 \cdot 8 \cdot 7 \cdot 6 = 2^4 \cdot 3^2 \cdot 7^2 = 84^2$.   So we get: $S = 84$.

Now, we have: $r = \frac{2S}{a+b+c}$ and $4SR = abc$.

So we can get first: $r = 2 \cdot 84/42 = 4$. And next, we get:

$R = abc/4S = 13 \cdot 14 \cdot 15/(4 \cdot 84) = 13 \cdot 2 \cdot 7 \cdot 3 \cdot 5/(2 \cdot 2 \cdot 7 \cdot 3 \cdot 4) = 13 \cdot 5/(2 \cdot 4) = 65/8$.

In short:

We have: $S = \sqrt{t(t-a)(t-b)(t-c)}$ where $2t = a + b + c$.

So first, we get: $2t = a + b + c = 13 + 14 + 15 = 42 \Rightarrow t = 21$.

Thus next, we can get:
$t(t-a)(t-b)(t-b) = 21(21-13)(21-14)(21-15) = 21 \cdot 8 \cdot 7 \cdot 6 = 2^4 \cdot 3^2 \cdot 7^2 = 84^2$.

So we get: $S = 84$.

And we have: $r = \frac{2S}{a+b+c}$ and $4SR = abc$.   So we get:

$r = 2 \cdot 84/42 = 4$, and $R = abc/4S = 13 \cdot 14 \cdot 15/(4 \cdot 84) = 13 \cdot 2 \cdot 7 \cdot 3 \cdot 5/(2 \cdot 2 \cdot 7 \cdot 3 \cdot 4) = 65/8$.

Suggestions or Solutions
To the **Problem** in the Example **3**

Find the area of a tetragon *ABCD* below assuming *B* = π/3, and *C* = 5π/12.

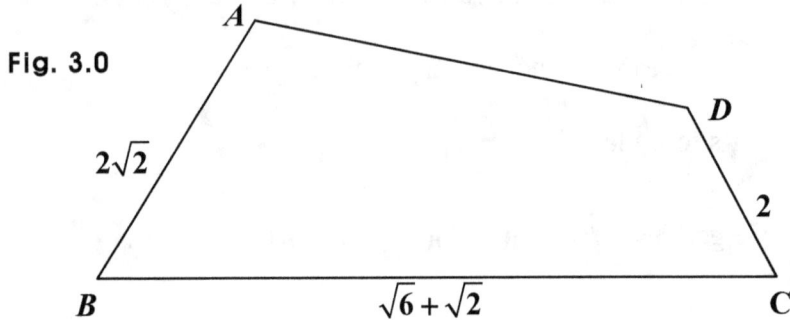

Fig. 3.0

A tetragon can be said to be a sum of many triangles, and thus, we can partition the tetragon into two triangles as shown below:

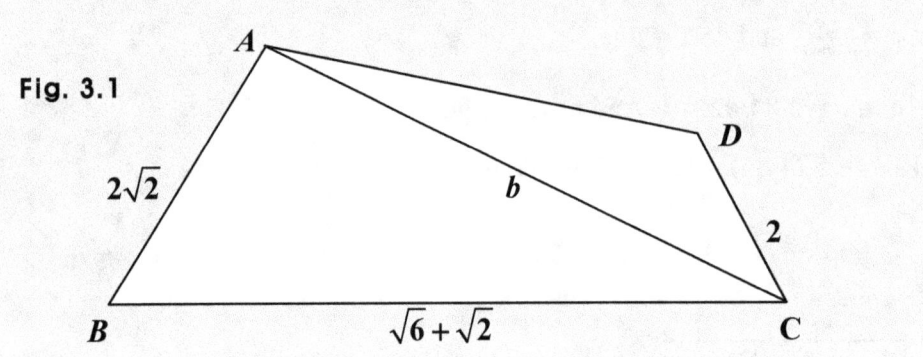

Fig. 3.1

So assuming *T* is the area of the tetragon, we can put *T* the way below:

$T = t_0 + t_1$, where t_0 is the area of $\triangle ABC$, and t_1 is the area of $\triangle ADC$.

How then, can we get the area t_0?

Knowing two sides and the angle between the two, we can get the area.

For instance, assuming *S* is the area of $\triangle XYZ$, we can get: $S = (xy \sin Z)/2$.

We know the angle **B** and the two sides **a** and **c** in **ΔABC**.

And we have: $B = \pi/3$, $a = 2\sqrt{2}$, and $c = \sqrt{6} + \sqrt{2}$. So we can get:

$$t_0 = (ca \sin B)/2 = \{(\sqrt{6} + \sqrt{2})2\sqrt{2} \sin 60°\}/2$$

$$= (\sqrt{6} + \sqrt{2})2\sqrt{2} \cdot \tfrac{\sqrt{3}}{2} \cdot \tfrac{1}{2} = (6 + 2\sqrt{3}) \cdot \tfrac{1}{2} = 3 + \sqrt{3}.$$

And the next is t_1, which is the area of **ΔADC**.

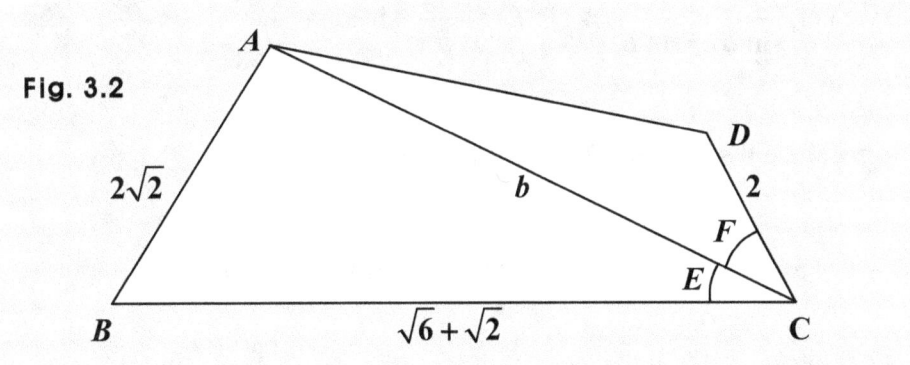

Fig. 3.2

Knowing two sides and the angle between the two, we can get the area.
We don't know however, the side **b** and the angle **F**.
We can get **b** though. How?

We have the cosine rule: $x^2 = y^2 + z^2 - 2yz \cos X$ where X is the angle between y an z.
So using the rule, we get:

$$b^2 = (2\sqrt{2})^2 + (\sqrt{6} + \sqrt{2})^2 - 2(2\sqrt{2})(\sqrt{6} + \sqrt{2}) \cdot \tfrac{1}{2}$$

$$= 8 + 8 + 4\sqrt{3} - 4\sqrt{3} - 4 = 12 \Rightarrow b = 2\sqrt{3}.$$

And thus, getting the angle **F**, we can get the area t_1, which is **(2b sin F)/2**.
How then, can we get the angle **F**?

We know in the triangle ABC, the angles B and C and the two sides, one is b, which is $2\sqrt{3}$, and the other is $2\sqrt{2}$ facing the angle E. So what?

Since we know the angle C, finding the angle E, we can get the angle F.

And finding the angle E, we can use the sine rule: $\dfrac{x}{\sin X} = \dfrac{y}{\sin Y} = \dfrac{z}{\sin Z} = 2R$ for $\triangle XYZ$.

So using the rule, we get: $\dfrac{b}{\sin B} = \dfrac{2\sqrt{2}}{\sin E} \Rightarrow \sin E = \dfrac{2\sqrt{2}}{b} \cdot \sin B = \dfrac{\sqrt{2}}{\sqrt{3}} \cdot \dfrac{\sqrt{3}}{2} = \dfrac{\sqrt{2}}{2}$.

Thus, we get: $E = \pi/4$ since $0 < E < 5\pi/12$.   And we know: $C = E + F$.

So we get: $F = C - E = 5\pi/12 - \pi/4 = \pi/6$.   And the area t_1 is $(2b \sin F)/2$.

Thus, we get: $t_1 = (2b \sin F)/2 = 2 \cdot 2\sqrt{3} \cdot \frac{1}{2} \cdot \frac{1}{2} = \sqrt{3}$.

And thus, the area of the tetragon $T = t_0 + t_1 = 3 + \sqrt{3} + \sqrt{3} = 3 + 2\sqrt{3}$.

8.0. **Trigonometry Dynamic 1**

In large, we can say there are two categories in trigonometry: static and dynamic. Static trigonometry begins with a right triangle, and stays with a right triangle.

So static trigonometry remains within a right triangle, which is normal or ordinary. Is there a right triangle though, not normal or ordinary?

Expanding the idea of a right triangle normal, we get into trigonometry that can be said to be dynamic. In the dynamic trigonometry, we can use right triangles not only normal but transcendental, too. What then, do we mean by a right triangle transcendental?

Normally in math, transcendental means non-algebraic. In this case though, it means supernatural or beyond common thought. So a right triangle transcendental can be said to be paranormal, and thus, is certainly not a normal triangle. Why not normal, though?

In a triangle normal, the length of any side cannot be non-positive. So if it's normal, the length of each side cannot be negative or 0. What then, about triangles transcendental?

A transcendental right triangle is still a right triangle, but can have a side, the length of which is not positive. More specifically, if the side is not the hypotenuse, the side can have a length negative or 0 as well as positive. So the length can have all kinds of values.

And thus, such a right triangle is said to be transcendental.

So in a right triangle transcendental, the adjacent and the opposite can be positive, 0, or negative. And thus, we can give any real number to the adjacent and the opposite.

Now, using a right triangle transcendental, we run dynamic trigonometry. So running trigonometry dynamic, we use a transcendental right triangle, where we can use any real number as the adjacent or the opposite. How then, do we get such a triangle?

Let's get back to the terminal ray that keeps turning about the origin in the *x-y* plane.

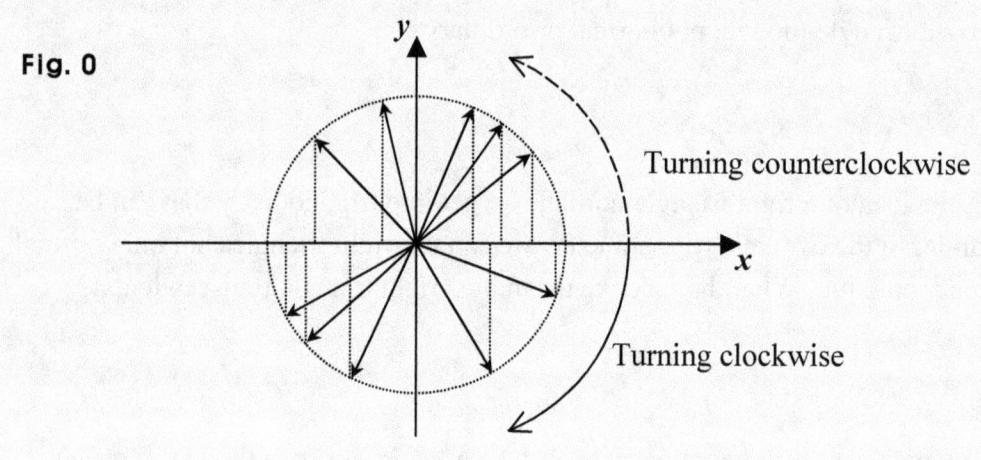

Fig. 0

Turning counterclockwise

Turning clockwise

Then, at every moment, taking the ray as a hypotenuse, we can see a lot of right triangles. In each triangle, we can use an *x*-coordinate as the adjacent. What then, is the opposite?

We can use a *y*-coordinate as (the value of) the opposite.
And the point that has the coordinates is the endpoint in a line segment used as the hypotenuse, where the other endpoint is at the origin. So the hypotenuse is the ray that keeps turning about the origin in the *x-y* plane.

And thus, the point that has the coordinates is the terminal point of the ray, that is, the arrowhead.

And we know that the length of the ray stays the same while the ray is turning about the origin. So the terminal point is making a circle, where the radius is the length of the ray, and the center is at the origin, of course.

And thus, assuming the ray is the hypotenuse in a right triangle, the *x*-coordinate at the terminal point is (the value of) the adjacent, and the *y*-coordinate is the opposite, we can see that being a right triangle, a right triangle keeps *changing* as the ray keeps turning.

So we are now working with a right triangle dynamic.

Now, when the ray is on either of the coordinate axes, or is in a quadrant other than the first one, it is not the case where both coordinates at the terminal point are positive.

So it is not the case where both the adjacent and the opposite are positive, which cannot happen in a normal right triangle. So the right triangle being made while the ray is turning can be said to be transcendental.

What then, is the purpose of the idea of a right triangle transcendental?

We know an angle is an amount of turning of the ray described above. So the ray turning makes angles. And if it turns *counterclockwise*, the angles are *positive*. If clockwise, it makes angles negative. And of course, no turning means that the angle is $0°$.

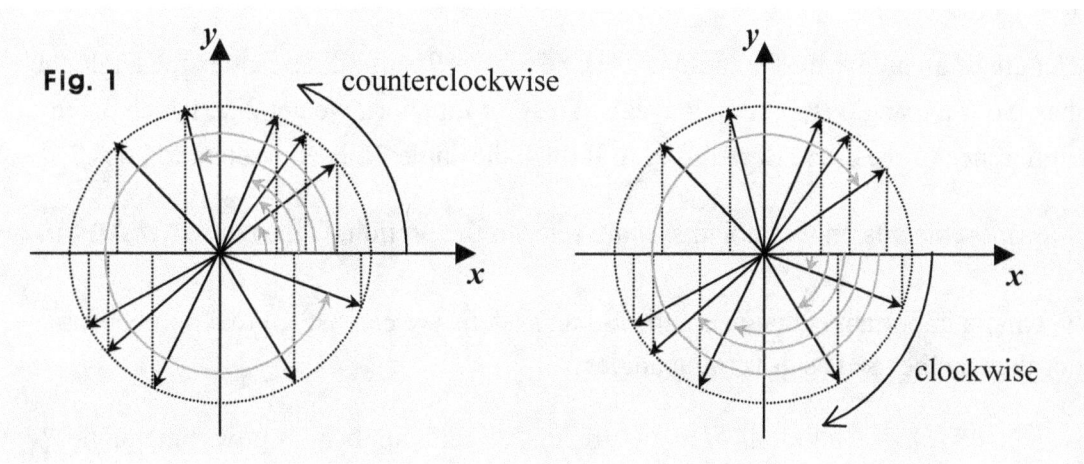

Now, we know a trig-ratio as **sin θ** has a governing angle, which is θ in the ratio **sin θ**.

So another major difference between the static and the dynamic is in governing angle, which is the purpose questioned above.
In static trigonometry, we work with triangles not transcendental but normal. And we cannot use all angles as governing angles.   Why not?

A governing angle is the angle between the adjacent and the hypotenuse in a right triangle. And one angle is $90°$ in the triangle, so the sum of the other two angles is $90°$. One of the other two is a governing angle. And thus, a governing angle is between $0°$ and $90°$ when we work in static trigonometry.

In dynamic trigonometry however, we can use as a governing angle any angle made by any turning of the ray. It can turn clockwise and counterclockwise. And also, before it turns, the angle is $0°$. So a governing angle can be $0°$ or any angle positive or negative. In other words, we can use all angles as governing angles for trig-ratios.   So what?

We can put all angles in a different system where we can specify angles.
The system is a metrology called *radian* system, where we use 2π as the basis, which is the circumference of a unit circle. Why is the basis though, such a circumference, 2π?

The length of an arc where the radius is r is: $r\theta$ where θ is in radian, and is the angle the arc has. So if the angle θ is 2π radian, called just 2π for short, we get $2\pi r$, which is the circumference of the circle of radius r. And thus, the angle 2π is equivalent to $360°$.

And for more details on the radian system, refer to the section, **Circles and Angles**.

Now, what's important is that using the radian system, we can use all real numbers as angles that can be used as governing angles.

How then, can we use all kinds of angles, that is, all real numbers as governing angles?

Let's now, get back again, to the terminal ray that keeps turning about the origin in the *x*-*y* plane.

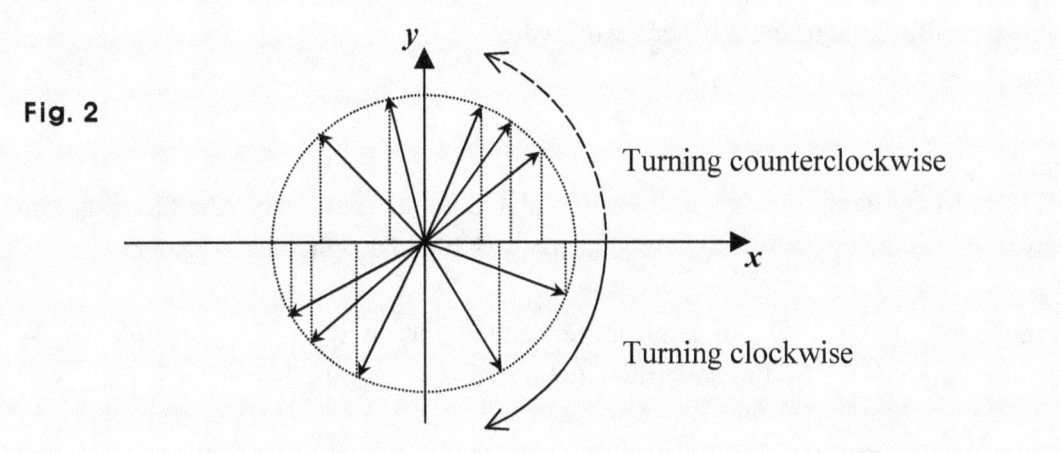

Fig. 2

Turning counterclockwise

Turning clockwise

Then, at every moment, taking the ray as a hypotenuse, we can see a lot of right triangles.

And in each triangle, we can use an *x*-coordinate as the adjacent, and use a *y*-coordinate as the opposite. And the point that has the coordinates is the terminal point of the ray, that is, the arrowhead.

And thus, assuming the ray is the hypotenuse in a right triangle, the *x*-coordinate at the terminal point is the adjacent, and the *y*-coordinate is the opposite, we can see that remaining a right triangle, a right triangle keeps changing as the ray keeps turning.

Now, when the ray is on either of the coordinate axes, or is in a quadrant other than the first one, it is not the case where both coordinates at the terminal point are positive.

So it is not the case where both the adjacent and the opposite are positive, which cannot happen in a normal right triangle.

So the right triangle being made while the ray is turning can be said to be transcendental. What then, is the governing angle in a right triangle transcendental?

In a right triangle normal, the adjacent and the hypotenuse make the governing angle. In a right triangle transcendental though, which is made in trigonometry dynamic, what makes the governing angle is the ray turning. And the dynamic trigonometry begins.

So to begin with, before the ray turns, it is assumed to be on the *x*-axis to the right of the origin, and the angle made by the ray is assumed to be 0.

What then, is the right triangle we can work with?

It is a right triangle transcendental, because the hypotenuse is the ray, and the adjacent is the ray, too, but the opposite is 0, which is not possible in a right triangle normal.

And in the trigonometry dynamic, the angle made by the ray turning is the angle we use as the governing angle. So the governing angle in this case is 0.

And we know the sine is: the opposite over the hypotenuse. So the sine is 0 because the opposite is 0 in this case. And thus, the sine of the governing angle 0, that is, **sin 0** is 0.

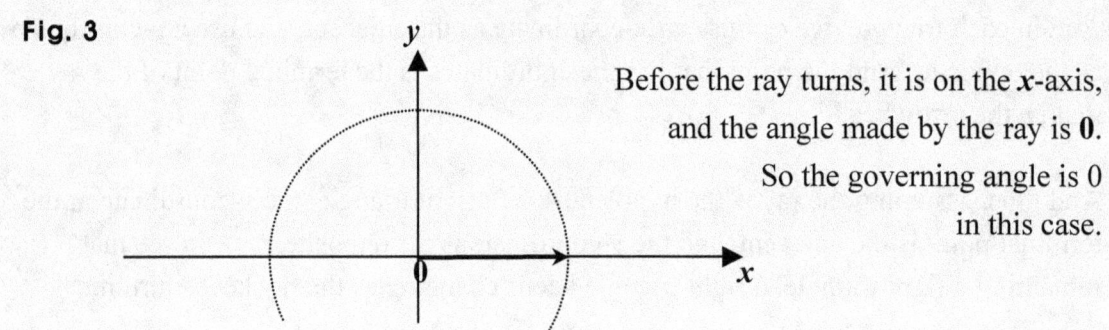

Fig. 3

Before the ray turns, it is on the *x*-axis, and the angle made by the ray is **0**. So the governing angle is 0 in this case.

And thus, though it does not look like any triangle at all, we can take it as a right triangle considering it as a right triangle transcendental.

What then, about the cosine?

The cosine is: the adjacent over the hypotenuse. So the cosine is 1 in the case above, because both are the same since each is the ray. And thus, the cosine of the governing angle 0, that is, **cos 0** is 1.

And next, what about the tangent?

The tangent is: the opposite over the adjacent. So the tangent is 0 in the case above, for the opposite is 0. And thus, the tangent of the governing angle 0, that is, **tan 0 is 0**.

- And next, when the ray is in the first quadrant, and thus, is between both coordinate axes, every triangle made can be said to be a right triangle normal.

Fig. 4

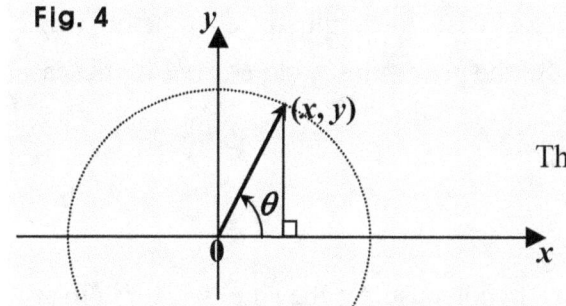

Suppose now, the length of the ray is 1, and the angle made by the ray is θ. Then, the governing angle is θ, and we can put the basic trig-ratios the way below:

Assuming (x, y) is the terminal point of the ray, we get:

$\sin \theta = y/1 = y$, $\cos \theta = x/1 = x$, and $\tan \theta = y/x$.

- Thus, we get: $\sin \theta = y$, $\cos \theta = x$, and $\tan \theta = y/x$.

And we have: $x^2 + y^2 = 1$.   So we get: $\underline{\sin^2 \theta + \cos^2 \theta = 1}$, which is a trig-identity. How do we get: $x^2 + y^2 = 1$, though?

We know the ray is of length 1, and is turning about the origin.

So if the ray makes a complete turn about the origin, the terminal point makes a circle of radius 1 centered at the origin, that is, a unit circle centered at the origin. And (x, y) is the terminal point, and thus, is in the unit circle. So the equation of such a unit circle is:

$x^2 + y^2 = 1$, which is the equation we can get using the distance formula.

Let's now, get back to the ray turning, and suppose the ray is on the y-axis, now. Then, the triangle we get is again, a right triangle transcendental.

That's because in the right triangle, the hypotenuse is the ray, and the opposite is the ray, too, but the adjacent is 0, which is not possible in a right triangle normal.

And we have another, which is possible in a right triangle not normal but transcendental. It is the governing angle, which is between 0 and π/2 in a right triangle normal, but does not have to be so in a right triangle transcendental.

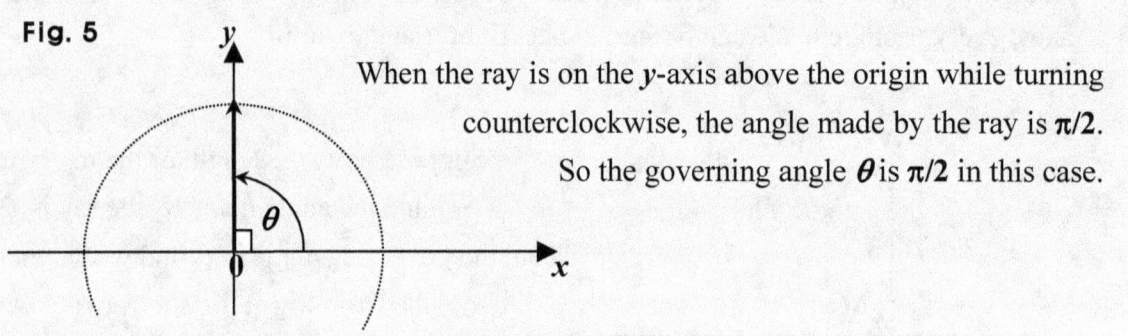

Fig. 5

When the ray is on the *y*-axis above the origin while turning counterclockwise, the angle made by the ray is **π/2**. So the governing angle *θ* is **π/2** in this case.

And we know the sine is: the opposite over the hypotenuse. So the sine is 1, for both are the same in this case. And thus, the sine of the governing angle π/2, that is, **sin π/2 is 1**.

What then, about the cosine?

We know the cosine is: the adjacent over the hypotenuse. So the cosine is 0, since the adjacent is 0. And thus, the cosine of the governing angle π/2, that is, **cos π/2 is 0**.

And next, what about the tangent?

We know the tangent is: the opposite over the adjacent. The adjacent is however, 0 in this case, and thus, causes the tangent to be undefined.   How come?

We know no denominator can be 0. So the tangent cannot be defined in this case. That is, **tan π/2 is not defined**. In other words, **tan π/2 does not exist**.

Usually though, it is said to be infinity. It's because turning counterclockwise, as the ray approaches the *y*-axis, the tangent of the governing angle gets bigger monotonically, and is huge, or rather, is as good as infinity when the ray gets very close to the *y*-axis.

We want to note that though, **tan π/2** does not exist, because the tangent cannot be defined for the angle of π/2, since the adjacent is 0, and is the denominator.

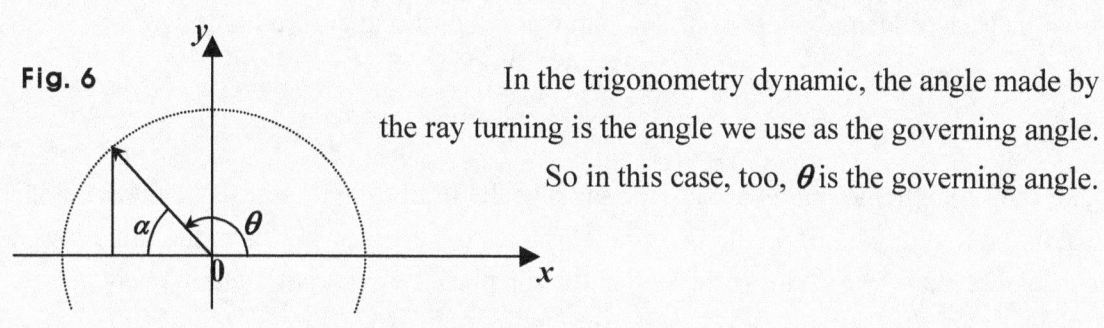

• And let's next, move on to the case where the terminal ray is in the second quadrant.

Fig. 6

In the trigonometry dynamic, the angle made by the ray turning is the angle we use as the governing angle. So in this case, too, *θ* is the governing angle.

Then, taking the ray as a hypotenuse, we can see a right triangle. And in the triangle, we can use an *x*-coordinate as the adjacent, and use a *y*-coordinate as the opposite. And the point that has the coordinates is the terminal point of the ray, that is, the arrowhead.

In this case though, the *x*-coordinate is negative, that is, the adjacent is negative.

Besides, the governing angle is not between 0 and π/2 but is greater than π/2.

So in this case, too, we get to work with a right triangle transcendental.

How in this case then, can we get trig-ratios?

To begin with, we want to note that in a right triangle transcendental, it is not the case where the governing angle is the angle between the adjacent and the hypotenuse.

As mentioned above, in the trigonometry dynamic, the angle we use as the governing angle is the angle made by the ray turning, about the origin, of course. So not the angle *α* above but the angle *θ* is the governing angle in the transcendental right triangle above.

Getting trig-ratios though, in a right triangle transcendental, too, we still get them the way we get them in a right triangle normal.

That is to say that in a right triangle transcendental, too, the sine is: the opposite over the hypotenuse, the cosine is: the adjacent over the hypotenuse, and the tangent is: the opposite over the adjacent. In other words, we need to make use of the angle α, too.

So getting trig-ratios in a right triangle transcendental, we work with not only the triangle transcendental but the one normal, too. What triangle normal though?

The right triangle normal is in fact, the same as the right triangle transcendental, but is put in the first quadrant. So the two triangles have different positions in the same **x-y** coordinate plane. And putting the two in the **x-y** plane, we can put them the way below:

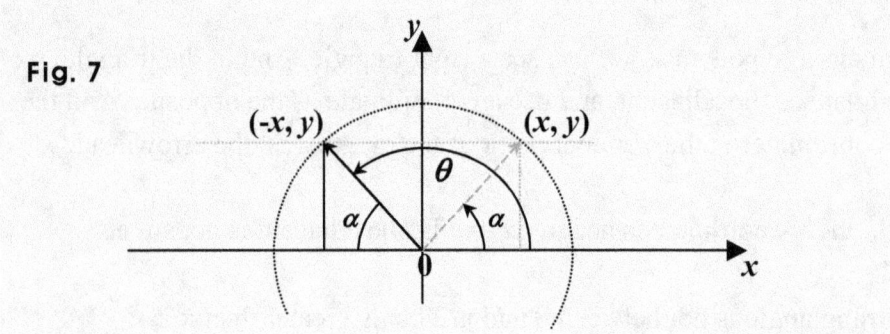

Fig. 7

Looking at the figure above, we can see that α is the governing angle in the triangle on the right hand side. And the triangle is the one normal mentioned above. And the angle α is the angle between the adjacent and the hypotenuse in the one transcendental, too. So both the normal and the transcendental are the same right triangles.

And thus, finding the trig-ratios in the triangle transcendental, we refer to the one normal.

Note however, we refer to the one normal. That is, we don't just use the trig-ratios in the one normal as the trig-ratios in the one transcendental. How then, can we get trig-ratios in the one transcendental?

Finding the coordinates of the terminal point in the ray, which is the hypotenuse in the one transcendental, we can get the trig-ratios in the one transcendental.   How come?

Using the one normal, we get first, the coordinates of the terminal point in the ray, which is the hypotenuse in the one transcendental, of course. We can get the coordinates using the fact that both the normal and the transcendental are the same. How can we get them?

Assuming the terminal point in the normal is (x, y), and using the fact above, we can see that the terminal point in the transcendental is $(-x, y)$. That is, in the transcendental, the adjacent is $-x$, and the opposite is y. Thus next, using $(-x, y)$, we can get trig-ratios in the transcendental.

So to begin with, we know the sine is: the opposite over the hypotenuse.

So assuming the hypotenuse is 1, we can see that the sine is y, for the opposite is y, and the hypotenuse is 1.

And thus, in the case where the ray is in the second quadrant, the sine of the governing angle θ, that is, **sin θ** is y, which equals **sin α**, too, since the angle α is the governing angle in the right triangle normal, and **sin α** is y.

So we get: **sin θ = sin α** in this case. And getting back to the figure above, we can see that $\theta = \pi - \alpha$. So we can say that if $\theta = \pi - \alpha$, we get: **sin θ = sin α**.

- That is to say that: **sin $(\pi - \alpha)$ = sin α**, which can be taken as a trig-identity.

What then, about the cosine?

We know the cosine is: the adjacent over the hypotenuse. So assuming again, the hypotenuse is 1, we can see that the cosine is $-x$, because the adjacent is $-x$.

And thus, in the case where the ray is in the second quadrant, the cosine of the governing angle θ, that is, **cos θ** is $-x$, which equals $-$**cos α**, too, since the angle α is the governing angle in the right triangle normal, and **cos α** is x. And we have: $\theta = \pi - \alpha$.

So we can see that if $\theta = \pi - \alpha$, we get: **cos θ = $-$cos α**.

- That is to say that: **cos (π – α) = –cos α**, which can be taken as another trig-identity.

And next, what about the tangent?

We know the tangent is: the opposite over the adjacent. So the tangent is –*y*/*x*, since the opposite is *y*, and the adjacent is –*x*.

And thus, in the case where the ray is in the second quadrant, the tangent of the governing angle *θ*, that is, **tan *θ*** is –*y*/*x*, which is **–tan α**, too, since the angle *α* is the governing angle in the right triangle normal, and **tan α** is *y*/*x*.

And we have: *θ* = π – *α*. So we can see that if *θ* = π – *α*, we get: **tan *θ* = –tan α**.

- That is to say that: **tan (π – α) = –tan α**, which can be taken as a trig-identity, too.

And in the next section, **Trigonometry Dynamic 2**, we are going to cover the case where the ray is on the *x*-axis, to the left of the origin, of course.

8.1. **Trigonometry Dynamic 2**

And next, when the ray is on the *x*-axis, to the left of the origin, of course, the triangle we get is again, a right triangle transcendental, too.

That's because in the right triangle we get, the hypotenuse is the ray, and the opposite is the ray, too, but the adjacent is 0, which is not possible in a right triangle normal.

And we have another, which is possible in a right triangle not normal but transcendental. It is the governing angle, which is between 0 and π/2 in a right triangle normal, but does not have to be so in a right triangle transcendental.

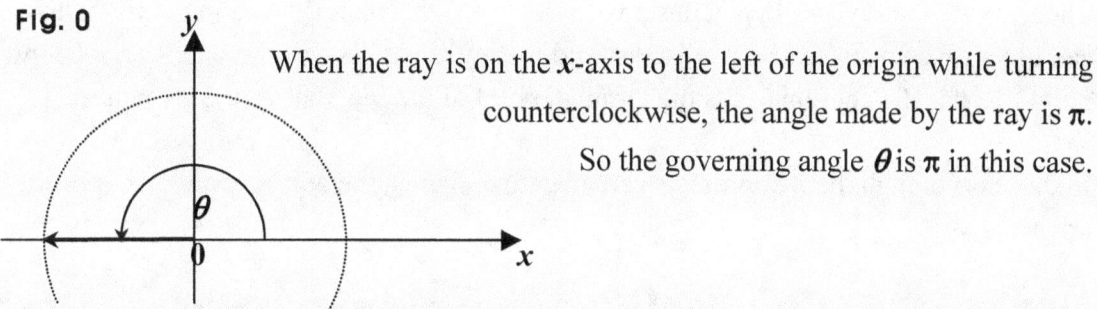

Fig. 0

When the ray is on the *x*-axis to the left of the origin while turning counterclockwise, the angle made by the ray is π.
So the governing angle *θ* is π in this case.

And we know the sine is: the opposite over the hypotenuse. So the sine is 0, for the opposite is 0 in this case. And thus, the sine of the governing angle π, that is, **sin π is 0**. What then, about the cosine?

We know the cosine is: the adjacent over the hypotenuse. So assuming again, the hypotenuse is 1, the cosine is -1, because we use as the adjacent the *x*-coordinate of the terminal point in the ray, and the *x*-coordinate is -1.

And thus, the cosine of the governing angle π, that is, **cos π is -1**.

And next, what about the tangent?

The tangent is: the opposite over the adjacent. So the tangent is 0 in the case above, for the opposite is 0. And thus, the tangent of the governing angle π, that is, **tan π is 0**.

- And let's next, move on to the case where the ray is in the *third quadrant*, and thus, is between both coordinate axes.

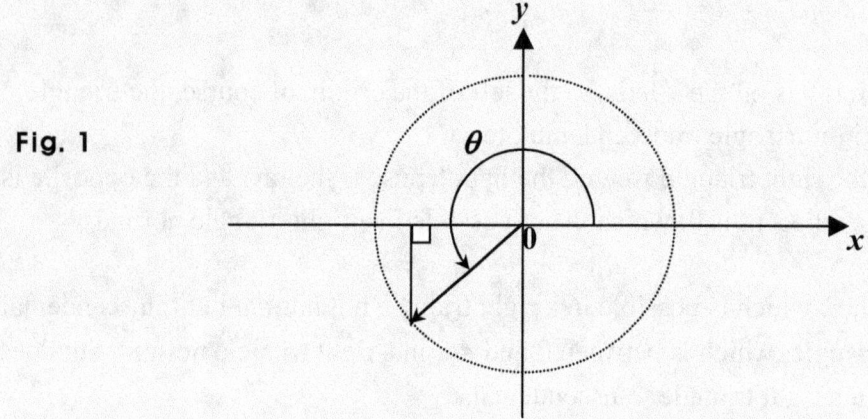

Fig. 1

Then, taking the ray as a hypotenuse, we can see a right triangle. And in the triangle, we can use an x-coordinate as the adjacent, and use a y-coordinate as the opposite. And the point that has the coordinates is the terminal point of the ray, that is, the arrowhead.

In this case though, both coordinates are negative, that is, the adjacent and the opposite are negative.

Besides, the governing angle is not between 0 and $\pi/2$ but is greater than $\pi/2$.
So in this case, too, we get to work with a right triangle transcendental.
How in this case then, can we get trig-ratios?

In the case where the ray is in the second quadrant, we have used the fact below:

Finding trig-ratios in a right triangle transcendental, we refer to the one normal, where the governing angle is the same as the angle between the adjacent and the hypotenuse in the one transcendental.

And referring to the one normal, we assume that the terminal point in the ray for the one normal is (**x, y**), and use the fact that both the normal and transcendental are the same.

Then, we can get the coordinates of the terminal point in the ray, which is the hypotenuse in the transcendental, of course.

That is to say that we can get the adjacent and the opposite in the triangle transcendental, because the **x**-coordinate of the terminal point in the adjacent, and the **y**-coordinate is the opposite.

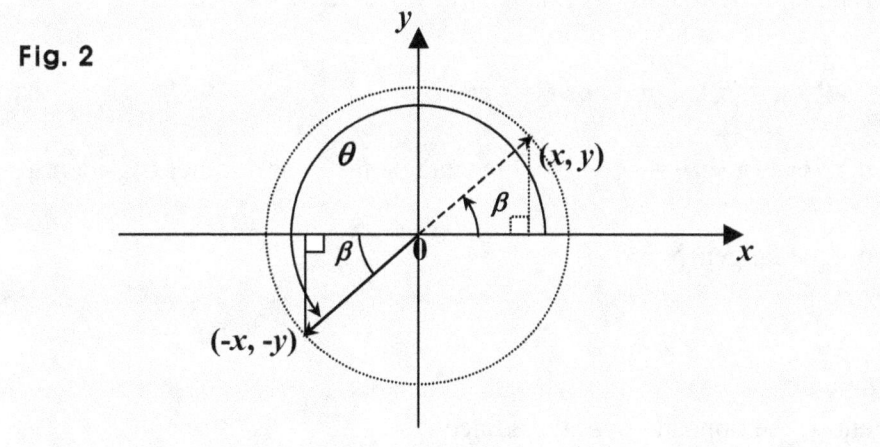

Fig. 2

And thus, assuming the terminal point in the ray for the normal is (**x, y**), we can see that (**-x, -y**) is the terminal point in the ray for the transcendental.

So to begin with, we know the sine is: the opposite over the hypotenuse.
So assuming the hypotenuse is 1, we can see that the sine is **–y**, for the opposite is **–y**, and the hypotenuse is 1.

And thus, in the case where the ray is in the third quadrant, the sine of the governing angle θ, that is, **sin θ** is **–y**, which equals **–sin β**, too, since the angle β is the governing angle in the right triangle normal, and **sin β** is **y**.

So we get: **sin θ = –sin β** in this case. And getting back to the figure above, we can see that $\theta = \pi + \beta$.   So we can say that if $\theta = \pi + \beta$, we get: **sin θ = –sin α**.

- That is to say that: **sin ($\pi + \beta$) = –sin β**, which can be taken as a trig-identity.

What then, about the cosine?

We know the cosine is: the adjacent over the hypotenuse. So assuming again, the hypotenuse is 1, we can see that the cosine is $-x$, because the adjacent is $-x$.

And thus, in the case where the ray is in the third quadrant, the cosine of the governing angle θ, that is, **cos θ** is $-x$, which equals $-\cos\beta$, too, since the angle β is the governing angle in the right triangle normal, and **cos β** is x. And we have: $\theta = \pi + \beta$.

So we can see that if $\theta = \pi + \beta$, we get: **cos $\theta = -\cos\beta$.**

- That is to say that: **cos $(\pi + \beta) = -\cos\beta$**, which can be taken as another trig-identity.

And next, what about the tangent?

We know the tangent is: the opposite over the adjacent.
So the tangent is y/x, since the opposite is y, and the adjacent is x.

And thus, in the case where the ray is in the third quadrant, the tangent of the governing angle θ, that is, **tan θ** is y/x, which is **tan β**, too, since the angle β is the governing angle in the right triangle normal, and **tan β** is y/x.

And we have: $\theta = \pi + \beta$. So we can see that if $\theta = \pi + \beta$, we get: **tan $\theta = \tan\beta$.**

- That is to say that: **tan $(\pi + \beta) = \tan\beta$**, which can be taken as a trig-identity, too.

And next, when the ray is on the y-axis, below the origin, of course, the triangle we get is again, a right triangle transcendental, too.

That's because in the right triangle we get, the hypotenuse is the ray, but the opposite is the negative of the length of the ray, and the adjacent is 0, which is not possible in a right triangle normal.

And we have another, which is possible in a right triangle not normal but transcendental.

It is the governing angle, which is between 0 and π/2 in a right triangle normal, but does not have to be so in a right triangle transcendental.

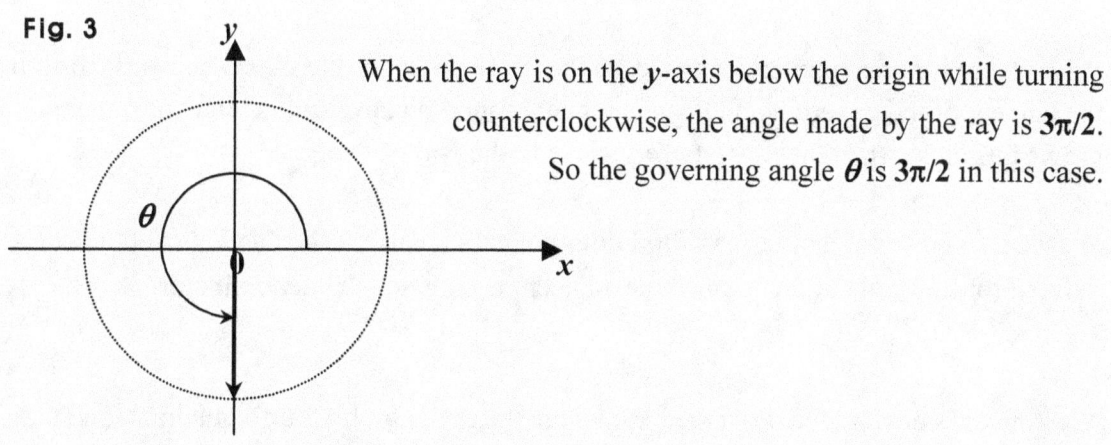

Fig. 3

When the ray is on the *y*-axis below the origin while turning counterclockwise, the angle made by the ray is **3π/2**. So the governing angle *θ* is **3π/2** in this case.

And we know the sine is: the opposite over the hypotenuse.

So assuming again, the hypotenuse is 1, the sine is –1.

That's because we use as the opposite the *y*-coordinate of the terminal point in the ray, and the *y*-coordinate is -1.

And thus, the sine of the governing angle 3π/2, that is, <u>**sin 3π/2** is -1</u>.

What then, about the cosine?

We know the cosine is: the adjacent over the hypotenuse. So the cosine is 0, since the adjacent is 0. And thus, the cosine of the governing angle 3π/2, that is, <u>**cos 3π/2** is 0</u>.

And next, what about the tangent?

We know the tangent is: the opposite over the adjacent. The adjacent is however, 0 in this case, and thus, causes the tangent to be undefined. How come?

We know no denominator can be 0.   So the tangent cannot be defined in this case.

That is, **tan 3π/2** is not defined. In other words, **tan 3π/2 does not exist**.

Usually though, as in the case of the tangent of π/2, **tan 3π/2 is said to be infinity**.

It's because turning counterclockwise, as the ray approaches the *y*-axis below the origin, the tangent of the governing angle gets bigger monotonically, and is huge, or rather, grows infinitely when the ray gets very close to the *y*-axis.

We want to note that though, **tan 3π/2** does not exist, because the tangent cannot be defined for the angle of 3π/2, since the adjacent is 0, and is the denominator.

• And let's next, move on to the case where the ray is in the fourth quadrant, and thus, is between both coordinate axes.

Fig. 4

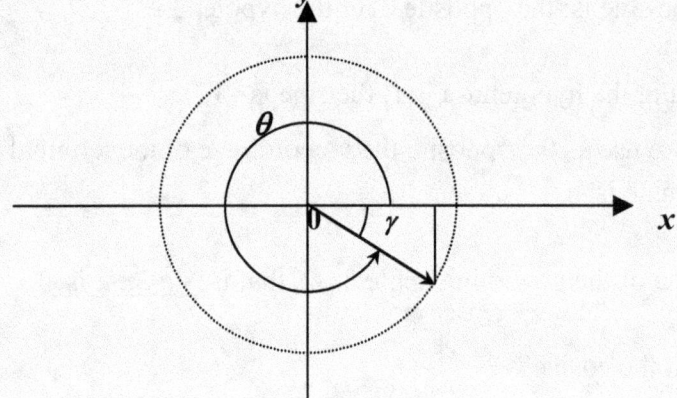

Then, taking the ray as a hypotenuse, we can see a right triangle. And in the triangle, we can use an *x*-coordinate as the adjacent, and use a *y*-coordinate as the opposite. And the point that has the coordinates is the terminal point of the ray, that is, the arrowhead.

In this case though, the *y*-coordinates is negative, that is, the opposite is negative.

Besides, the governing angle is not between 0 and π/2 but is greater than π/2.

So in this case, too, we get to work with a right triangle transcendental.
How in this case then, can we get trig-ratios?

In the case where the ray is in a quadrant other than the first, we used the fact below:

Finding trig-ratios in a right triangle transcendental, we refer to the one normal, where the governing angle is the same as the angle between the adjacent and the hypotenuse in the one transcendental.

And referring to the one normal, we assume that the terminal point in the ray for the one normal is (x, y), and use the fact that both the normal and transcendental are the same.

Then, we can get the coordinates of the terminal point in the ray, which is the hypotenuse in the transcendental, of course.

That is to say that we can get the adjacent and the opposite in the triangle transcendental, because the x-coordinate of the terminal point in the adjacent, and the y-coordinate is the opposite.

Fig. 5

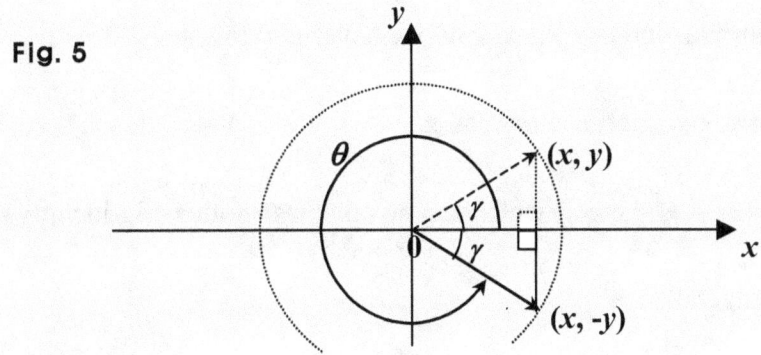

And thus, assuming the terminal point in the ray for the normal is (x, y), we can see that $(x, -y)$ is the terminal point in the ray for the transcendental.

So to begin with, we know the sine is: the opposite over the hypotenuse.

So assuming the hypotenuse is 1, we can see that the sine is $-y$, for the opposite is $-y$, and the hypotenuse is 1

And thus, in the case where the ray is in the fourth quadrant, the sine of the governing angle θ, that is, **sin θ** is $-y$, which equals $-\sin \gamma$, too, since the angle γ is the governing angle in the right triangle normal, and **sin γ** is y.

So we get: $\sin \theta = -\sin \gamma$ in this case. And getting back to the figure above, we can see that $\theta = 2\pi - \gamma$.

So we can say that if $\theta = 2\pi - \gamma$, we get: $\sin \theta = -\sin \gamma$.

- That is to say that: $\sin (2\pi - \gamma) = -\sin \gamma$, which can be taken as a trig-identity.

What then, about the cosine?

We know the cosine is: the adjacent over the hypotenuse.
So assuming again, the hypotenuse is 1, we can see that the cosine is x, because the adjacent is x.

And thus, in the case where the ray is in the fourth quadrant, the cosine of the governing angle θ, that is, $\cos \theta$ is x, which equals $\cos \gamma$, too, since the angle γ is the governing angle in the right triangle normal, and $\cos \gamma$ is x. And we have: $\theta = 2\pi - \gamma$.

So we can see that if $\theta = 2\pi - \gamma$, we get: $\cos \theta = \cos \gamma$.

- That is to say that: $\cos (2\pi - \gamma) = \cos \gamma$, which can be taken as another trig-identity.

And next, what about the tangent?

We know the tangent is: the opposite over the adjacent.

So the tangent is $-y/x$, since the opposite is $-y$, and the adjacent is x.

And thus, in the case where the ray is in the fourth quadrant, the tangent of the governing angle θ, that is, $\tan \theta$ is $-y/x$, which is $-\tan \gamma$, too, since the angle γ is the governing angle in the right triangle normal, and $\tan \gamma$ is y/x.

And we have: $\theta = 2\pi - \gamma$. So we can see that if $\theta = 2\pi - \gamma$, we get: $\tan \theta = -\tan \gamma$.

- That is to say that: $\tan (2\pi - \gamma) = -\tan \gamma$, which can be taken as a trig-identity, too.

8.2. **Trigonometry Dynamic 3**

And let's now, put threads together. Suppose first, the length of the ray turning is 1, and θ is the angle made by the ray turning.

Then, to begin with, before the ray turns, so when the ray is initially on the *x*-axis to the right of the origin, and thus, when $\theta = 0$, we get: **sin** $\theta = 0$, **cos** $\theta = 1$, and **tan** $\theta = 0$.

Fig. 0

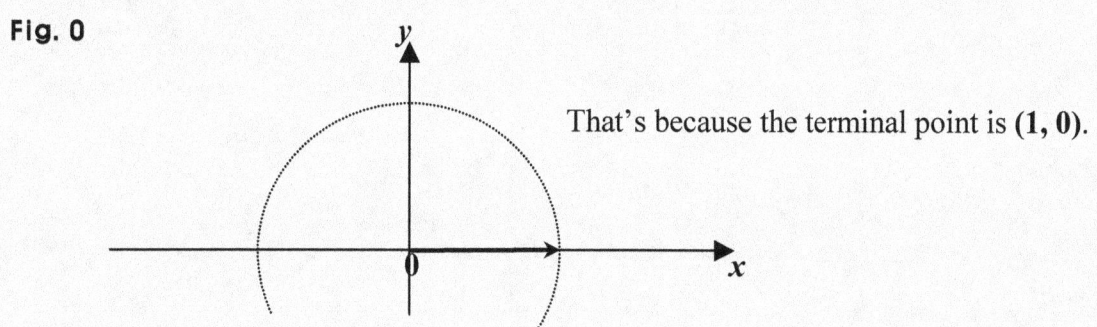

That's because the terminal point is **(1, 0)**.

Next, when the ray is in the first quadrant, that is, $0 < \theta < \pi/2$, we get:
sin $\theta = y$, **cos** $\theta = x$, and **tan** $\theta = y/x$, assuming the terminal point is **(x, y)**.

Fig. 1

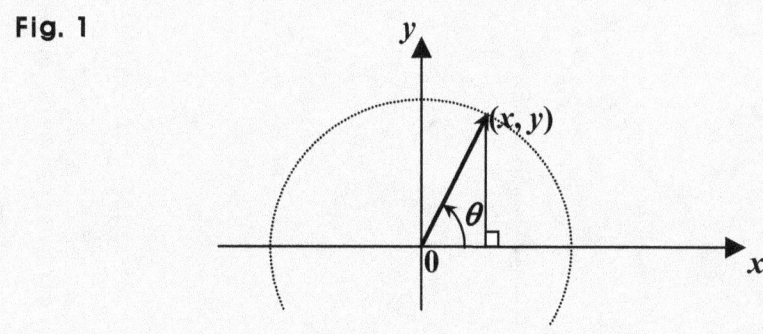

Next, when the ray is on the *y*-axis above the origin, that is, $\theta = \pi/2$, we get: **sin $\theta = 1$**, and **cos $\theta = 0$**, but no **tan θ**, for the adjacent is 0, so **tan θ** is not defined for $\theta = \pi/2$.

Fig. 2

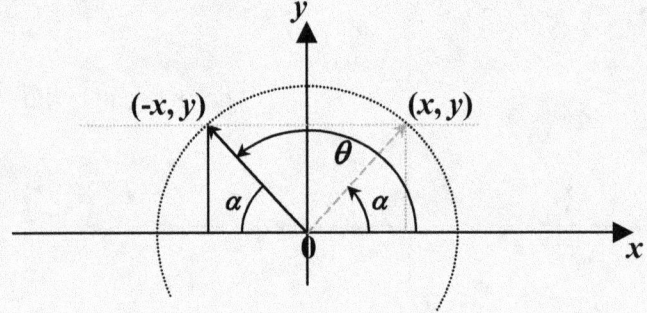

That's because the terminal point is **(0, 1)** in this case.

Next, when the ray is in the second quadrant, that is, $\pi/2 < \theta < \pi$, we get: **sin $\theta = y$**, **cos $\theta = -x$**, and **tan $\theta = -y/x$**, since the terminal point in the ray is **(-x, y)**.

Fig. 3

Next, when the ray is on the *x*-axis to the left of the origin, that is, $\theta = \pi$, we get: **sin $\theta = 0$**, **cos $\theta = -1$**, and **tan $\theta = 0$**, since the terminal point in the ray is **(-1, 0)**.

Fig. 4

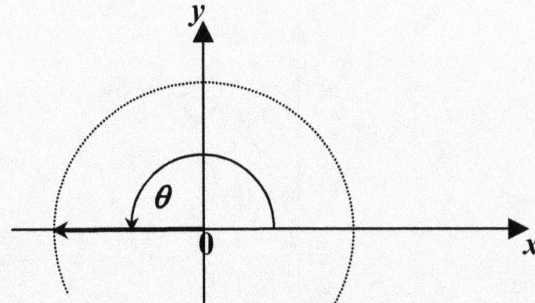

Next, when the ray is in the third quadrant, that is, **π < θ < 3π/2**, we get:

sin θ = -*y*, **cos θ** = -*x*, and **tan θ** = *y/x*, since the terminal point in the ray is (**-x, -y**).

Fig. 5

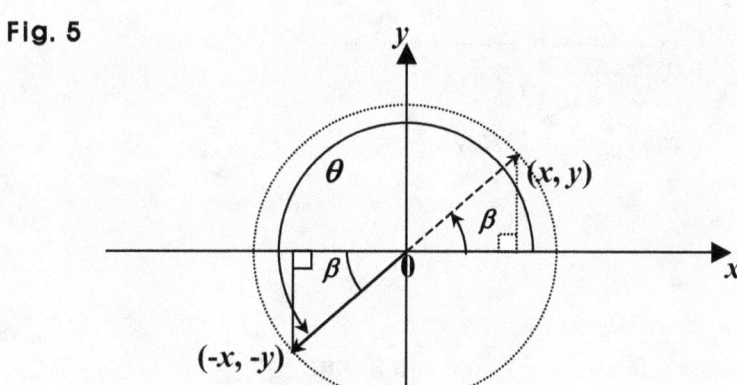

Next, when the ray is on the *y*-axis below the origin, that is, **θ = 3π/2**, we get: **sin θ = -1**, and **cos θ = 0**, but no **tan θ**, for the adjacent is 0, so **tan θ** is not defined for **θ = 3π/2**..

Fig. 6

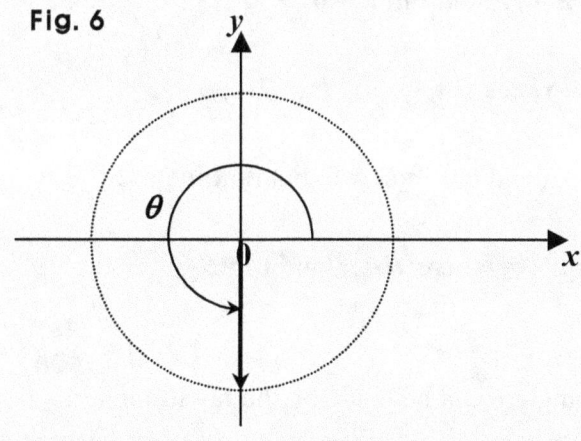

And next, when the ray is in the fourth quadrant, that is, **3π/2 < θ < 2π**, we get:

sin θ = -*y*, **cos θ** = *x*, and **tan θ** = -*y/x*, since the terminal point in the ray is (**x, -y**).

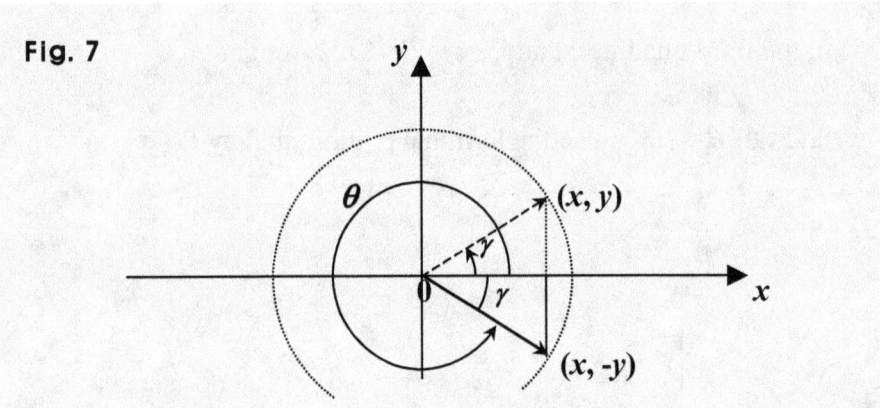

Fig. 7

So in sum:

Fist, when $\theta = 0$, we get: **sin 0 = 0, cos 0 = 1**, and **tan 0 = 0.**

Next, when $0 < \theta < \pi/2$, we get: **sin $\theta = y$, cos $\theta = x$**, and **tan $\theta = y/x$.**

Next, when $\theta = \pi/2$, we get: **sin π/2 = 1**, and **cos π/2 = 0**, but no **tan π/2.**

Next, when $\pi/2 < \theta < \pi$, we get: **sin $\theta = y$, cos $\theta = -x$**, and **tan $\theta = -y/x$.**

Next, when $\theta = \pi$, we get: **sin π = 0, cos π = -1**, and **tan π = 0.**

Next, when $\pi < \theta < 3\pi/2$, we get: **sin $\theta = -y$, cos $\theta = -x$**, and **tan $\theta = y/x$.**

Next, when $\theta = 3\pi/2$, we get: **sin 3π/2 = -1**, and **cos 3π/2 = 0**, but no **tan 3π/2.**

And next, when $3\pi/2 < \theta < 2\pi$, we get: **sin $\theta = -y$, cos $\theta = x$**, and **tan $\theta = -y/x$.**

So we have covered eight cases where the angle θ can be made by the ray turning. Now, the ray is placed back on the x-axis. What then, is the angle θ?

The angle θ is 2π, because the ray has just made a complete turn. What will happen then, to the trig-ratios if the ray keeps turning, counterclockwise, of course?

The eight cases described above will repeat themselves, in the same sequence, of course.

To begin with, when $\theta = 0$, we get: **sin 0 = 0**, **cos 0 = 1**, and **tan 0 = 0**.

So when $\theta = 2\pi$, we will get this, too: **sin 2π = 0**, **cos 2π = 1**, and **tan 2π = 0**.

And thus, for any integer $n \geq 0$, when $\theta = 2n\pi$, we will get this, also:

sin θ = 0, **cos θ = 1**, and **tan θ = 0**.

Suppose next, $0 < \alpha < \pi/2$, that is, α is an acute angle, and n is an integer ≥ 0. Then:

First, when $0 < \theta < \pi/2$, we get: **sin θ = y**, **cos θ = x**, and **tan θ = y/x**.

So when $\theta = 2\pi + \alpha$, we will get this, too: **sin θ = y**, **cos θ = x**, and **tan θ = y/x**.

And thus, when $\theta = 2n\pi + \alpha$, we will get this, also: **sin θ = y**, **cos θ = x**, and **tan θ = y/x**.

Next, when $\theta = \pi/2$, we get: **sin $\pi/2$ = 1**, and **cos $\pi/2$ = 0**, but no **tan $\pi/2$**.

So when $\theta = 2\pi + \pi/2$, we will get this, also: **sin θ = 1**, **cos θ = 0**, but no **tan θ**.

And thus, when $\theta = 2n\pi + \pi/2$, we will get this, too: **sin θ = 1**, **cos θ = 0**, but no **tan θ**.

Next, when $\pi/2 < \theta < \pi$, we get: **sin θ = y**, **cos θ = -x**, and **tan θ = -y/x**.

And we have: $0 < \alpha < \pi/2 \Rightarrow 0 > -\alpha > -\pi/2 \Rightarrow \pi/2 < \pi - \alpha < \pi$. So we can set: $\theta = \pi - \alpha$.

Thus, when $\theta = \pi - \alpha$, we get: **sin θ = y**, **cos θ = -x**, and **tan θ = -y/x**.

So when $\theta = 2\pi + \pi - \alpha$, we will get this, too: **sin θ = y**, **cos θ = -x**, and **tan θ = -y/x**.

And we can have: $\theta = 2n\pi + \pi - \alpha = (2n + 1)\pi - \alpha$, where α is an acute angle.

So when $\theta = (2n + 1)\pi - \alpha$, we will get this, too: **sin θ = y**, **cos θ = -x**, and **tan θ = -y/x**.

Next, when $\theta = \pi$, we get: **sin π = 0**, **cos π = -1**, and **tan π = 0**.

So when $\theta = 2\pi + \pi$, we will get this, also: **sin π = 0**, **cos π = -1**, and **tan π = 0**. And thus, when $\theta = 2n\pi + \pi = (2n + 1)\pi$, we will get this, too: **sin π = 0**, **cos π = -1**, and **tan π = 0**.

Next, when $\pi < \theta < 3\pi/2$, we get: **sin θ = -y**, **cos θ = -x**, and **tan θ = y/x**.

And we have: **0 < α < $\pi/2$ $\Rightarrow$ π < π + α < 3π/2**. So we can set: $\theta = \pi + \alpha$.

Thus, when $\theta = \pi + \alpha$, we get: **sin θ = -y**, **cos θ = -x**, and **tan θ = y/x**.

So when $\theta = 2\pi + \pi + \alpha$, we will get this, too: **sin θ = -y**, **cos θ = -x**, and **tan θ = y/x**.

And we can have: $\theta = 2n\pi + \pi + \alpha = (2n + 1)\pi + \alpha$, where α is an acute angle.

So when $\theta = (2n + 1)\pi + \alpha$, we will get this, also: **sin θ = -y**, **cos θ = -x**, and **tan θ = y/x**.

Next, when $\theta = 3\pi/2$, we get: **sin 3π/2 = -1**, and **cos 3π/2 = 0**, but no **tan 3π/2**.

So when $\theta = 2\pi + 3\pi/2$, we will get this, too: **sin θ = -1**, and **cos θ = 0**, but no **tan θ**.

And we can have: $\theta = 2n\pi + 3\pi/2 = 2n\pi + \pi + \pi/2 = (2n + 1)\pi + \pi/2$.

So when $\theta = 2n\pi + 3\pi/2$ or $\theta = (2n + 1)\pi + \pi/2$, we will get this, also: **sin θ = -1**, and **cos θ = 0**, but no **tan θ**.

And next, when $3\pi/2 < \theta < 2\pi$, we get: **sin θ = -y**, **cos θ = x**, and **tan θ = -y/x**.   And we have: **0 < α < $\pi/2$ $\Rightarrow$ 0 > -α > -$\pi/2$ $\Rightarrow$ 3π/2 < 2π - α < 2π**.   So we can set: $\theta = 2\pi - \alpha$.

And we can have: $\theta = 2\pi + 2\pi - \alpha = 4\pi - \alpha$.

So when $\theta = 4\pi - \alpha$, we get this, too: **sin θ = -y**, **cos θ = x**, and **tan θ = -y/x**.

And also, we can have: $\theta = 2n\pi + 2\pi - \alpha = 2(n + 1)\pi - \alpha$, where α is an acute angle.

So when $\theta = 2(n + 1)\pi - \alpha$, we get this, too: $\sin \theta = -y$, $\cos \theta = x$, and $\tan \theta = -y/x$.

And we can put it this way, too:

We know n is an integer ≥ 0. So $2(n + 1)$ is an even integer ≥ 2.

So we can set: $\theta = 2k\pi - \alpha$, where k is an integer ≥ 1, and α is an acute angle.

And thus, when $\theta = 2k\pi - \alpha$, we get this, too: $\sin \theta = -y$, $\cos \theta = x$, and $\tan \theta = -y/x$.

So putting the eight cases in general, we can put them the way below:

Suppose first, $0 < \alpha < \pi/2$, that is, α is an acute angle, and n is an integer ≥ 0.   Then:

To begin with, when $\theta = 2n\pi$, we get: $\sin \theta = 0$, $\cos \theta = 1$, and $\tan \theta = 0$.

Next, when $\theta = 2n\pi + \alpha$, we get: $\sin \theta = y$, $\cos \theta = x$, and $\tan \theta = y/x$.

Next, when $\theta = 2n\pi + \pi/2$, we get: $\sin \theta = 1$, $\cos \theta = 0$, but no $\tan \theta$.

Next, when $\theta = (2n + 1)\pi - \alpha$, we get: $\sin \theta = y$, $\cos \theta = -x$, and $\tan \theta = -y/x$.

Next, when $\theta = (2n + 1)\pi$, we get: $\sin \pi = 0$, $\cos \pi = -1$, and $\tan \pi = 0$.

Next, when $\theta = (2n + 1)\pi + \alpha$, we get: $\sin \theta = -y$, $\cos \theta = -x$, and $\tan \theta = y/x$.

Next, when $\theta = (2n + 1)\pi + \pi/2$, we get: $\sin \theta = -1$, and $\cos \theta = 0$, but no $\tan \theta$.

And next, when $\theta = 2(n + 1)\pi - \alpha$, or $\theta = 2k\pi - \alpha$, where k is an integer ≥ 1, we get: $\sin \theta = -y$, $\cos \theta = x$, and $\tan \theta = -y/x$.

So do we now have to memorize them all?

Memorizations can help, of course. And learning things, we need to memorize some of those. It is often the case however, memories alone do not solve problems. What does do then, solve problems?

Reasoning does, together with understanding. And understanding the material we learn, we can often see it getting into memory quite quickly, too. And in fact, understanding the material above, and reasoning properly, we don't really have to memorize much of it.

Suppose for instance, we want to get the value of **sin 210°**.

We know: **sin 30° = 1/2**, and 210 = 180 + 30.   So quickly drawing a graph, we can get:

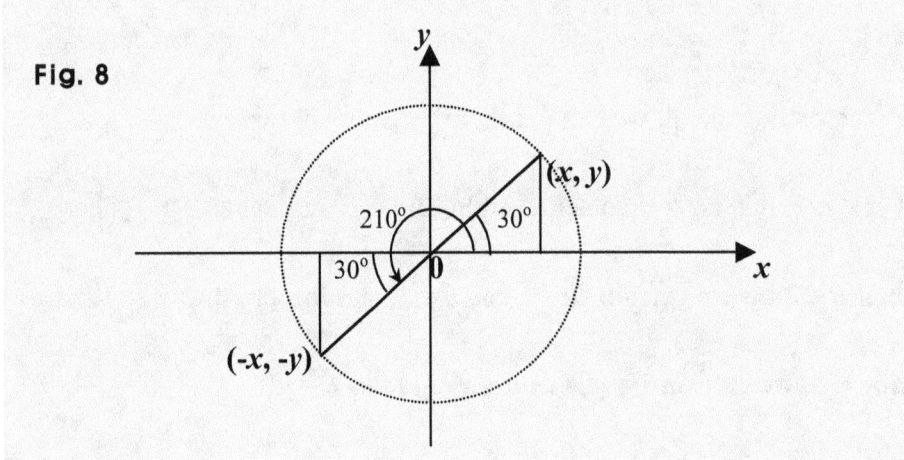

Fig. 8

Now, we know that the sine is: the opposite over the hypotenuse, and that *y* is the opposite in the right triangle above the *x*-axis.

So assuming the hypotenuse is 1, we get: $y = 1/2$.

And we have: **sin 210°** = -*y*, since –*y* is the opposite, and 1 is the hypotenuse in the right triangle below the *x*-axis.

So we get: **sin 210°** = -*y*/1 = -*y* = -1/2.

And also, comparing some of the eight cases above, we can get some trig-identities, too.

To begin with, when $\theta = 0$, we get: **sin 0 = 0, cos 0 = 1**, and **tan 0 = 0**.

And when $\theta = \pi$, we get: **sin π = 0, cos π = -1**, and **tan π = 0**.

So assuming *n* is an integer ≥ 0, we get: **sin $n\pi$ = 0, cos $n\pi$ = $(-1)^n$**, and **tan $n\pi$ = 0**.

Next, when $\theta = \pi/2$, we get: **sin $\pi/2$ = 1**, and **cos $\pi/2$ = 0**, but no **tan $\pi/2$**.

And when $\theta = 3\pi/2$, we get: **sin $3\pi/2$ = -1**, and **cos $3\pi/2$ = 0**, but no **tan $3\pi/2$**.

So assuming n is an integer ≥ 0, we get: **sin $(2n + 1)\pi/2 = (-1)^n$**, **cos $(2n + 1)\pi/2 = 0$**, but no **tan $(2n + 1)\pi/2$**.

• Next, when $0 < \theta < \pi/2$, we get: **sin $\theta = y$**, **cos $\theta = x$**, and **tan $\theta = y/x$**.

And when $\pi/2 < \theta < \pi$, we get: **sin $\theta = y$**, **cos $\theta = -x$**, and **tan $\theta = -y/x$**.

And we have: $0 < \alpha < \pi/2 \Rightarrow 0 > -\alpha > -\pi/2 \Rightarrow \pi/2 < \pi - \alpha < \pi$.

So we can put the two cases above the way below:

When $0 < \alpha < \pi/2$, we get: **sin $\alpha = y$**, **cos $\alpha = x$**, and **tan $\alpha = y/x$**.

When $\pi/2 < \pi - \alpha < \pi$, we get: **sin $(\pi - \alpha) = y$**, **cos $(\pi - \alpha) = -x$**, and **tan $(\pi - \alpha) = -y/x$**.

Thus, we get: **sin $(\pi - \alpha)$ = sin α**, **cos $(\pi - \alpha)$ = -cos α**, **tan $(\pi - \alpha)$ = -tan α**.

• Next, when $0 < \theta < \pi/2$, we get: **sin $\theta = y$**, **cos $\theta = x$**, and **tan $\theta = y/x$**.

And when $\pi < \theta < 3\pi/2$, we get: **sin $\theta = -y$**, **cos $\theta = -x$**, and **tan $\theta = y/x$**.

And we have: $0 < \alpha < \pi/2 \Rightarrow \pi < \pi + \alpha < 3\pi/2$.

So we can put the two cases above the way below:

When $0 < \alpha < \pi/2$, we get: **sin $\alpha = y$**, **cos $\alpha = x$**, and **tan $\alpha = y/x$**.

When $\pi < \pi + \alpha < 3\pi/2$, we get: **sin $(\pi + \alpha) = -y$**, **cos $(\pi + \alpha) = -x$**, and **tan $(\pi + \alpha) = y/x$**.

Thus, we get: **sin $(\pi + \alpha)$ = -sin α**, **cos $(\pi + \alpha)$ = -cos α**, and **tan $(\pi + \alpha)$ = tan α**.

• Next, when $0 < \theta < \pi/2$, we get: $\sin \theta = y$, $\cos \theta = x$, and $\tan \theta = y/x$.

And when $3\pi/2 < \theta < 2\pi$, we get: $\sin \theta = -y$, $\cos \theta = x$, and $\tan \theta = -y/x$.

And we have: $0 < \alpha < \pi/2 \Rightarrow 0 > -\alpha > -\pi/2 \Rightarrow 3\pi/2 < 2\pi - \alpha < 2\pi$.

So we can put the two cases above the way below:

When $0 < \alpha < \pi/2$, we get: $\sin \alpha = y$, $\cos \alpha = x$, and $\tan \alpha = y/x$.

If $\frac{3\pi}{2} < 2\pi - \alpha < 2\pi$, we get: $\sin (2\pi - \alpha) = -y$, $\cos (2\pi - \alpha) = x$, and $\tan (2\pi - \alpha) = -y/x$.

Thus, we get: $\sin (2\pi - \alpha) = -\sin \alpha$, $\cos (2\pi - \alpha) = \cos \alpha$, and $\tan (2\pi - \alpha) = -\tan \alpha$.

And setting the hypotenuse = 1, we have: $x^2 + y^2 = 1$. So we get: $\sin^2 \theta + \cos^2 \theta = 1$, which is a trig-identity very often used. What if however, the hypotenuse is not 1?

The identity still works. Assuming for instance, the hypotenuse is H, and the terminal point is (x, y), we get a circle of radius H, where (x, y) is an arbitrary point.

So we get: $x^2 + y^2 = H^2 \Rightarrow (\frac{x}{H})^2 + (\frac{y}{H})^2 = 1$. And we have: $\sin \theta = \frac{y}{H}$, and $\cos \theta = \frac{x}{H}$.

So we get: $\sin^2 \theta + \cos^2 \theta = 1$.

And the same is true for the case where the ray turns clockwise, too.

The only difference is in sign of each angle made by the ray turning. That is, if the ray turns clockwise, the angle made is negative, so the governing angle is negative.

However, all the identities above still work. That is, they are all satisfied with the governing angles negative, too. We can have identities though, regarding signs of angles.

And the identities are as follows: **sin (-θ) = -sin θ, cos (-θ) = cos θ**, and **tan (-θ) = tan θ**. How come?

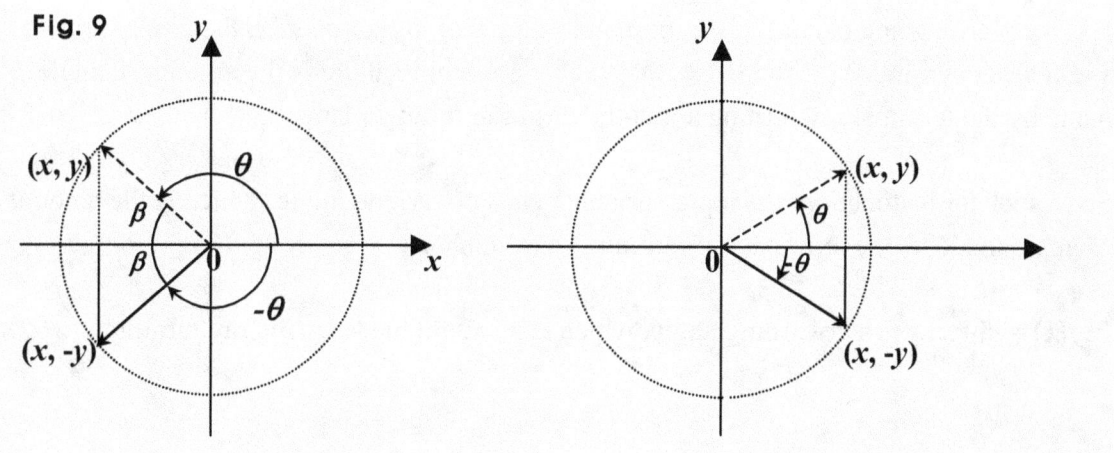

Fig. 9

No matter what right triangle it may be, the sine is: the opposite over the hypotenuse, the cosine is: the adjacent over the hypotenuse, and the tangent is: the opposite over the adjacent. So assuming the hypotenuse is 1, in each of the two figures above, we get:

sin θ = y, sin (-θ) = -y, cos θ = x, cos (-θ) = x, tan θ = y/x, and **tan (-θ) = -y/x.**

And thus, we get: **sin (-θ) = -sin θ, cos (-θ) = cos θ**, and **tan (-θ) = tan θ.**

Now, what's more important is that working in trigonometry dynamic, we can come up with another set of tools, called *trigonometric functions*, called *trig-functions* for short.

If not quite sure of functions, refer to **ALGEBRA EXAMPLES BASIC FUNCTIONS**.

A trig-function is a function of an angle, which is a governing angle, and thus, takes an angle as an input.

So for instance, assuming *f* is such a function, and *θ* is the input variable, we can set:

f(θ) = sin *θ*.

And of course, we can use as the variable any of other letters as *x* and *t*. So for instance, we can set: $f(x) = \sin x$, where *x* stands for an angle. What angle though?

It is a governing angle. And it is important to note that in $f(x) = \sin x$, the input variable *x* takes a governing angle, and also, that such a governing angle is the amount of angle made by the terminal ray turning about the origin in the *x-y* plane.

So in such a trig-function, the input variable gets a governing angle, which is the amount of angle made by the terminal ray turning. Thus, in short:

In $f(x) = \sin x$, *x* is a governing angle, which is an angle made by the ray turning.

What then, about outputs?

Each output is a number. What number though?

It is a trig-ratio, because *x* is a governing angle, and $\sin x$ is a trig-ratio.

That is, for each value of *x*, $f(x)$ is a trig-ratio. In short, *f* gets a trig-ratio.

So in sum, all the values of *x* are governing angles, and all the values of *f* are trig-ratios.

In short, every value of *x* is an angle, and every value of *f* is a ratio, which is a number.

Usually though, we use numbers as values of *x*, too.

That's because we often use as inputs, angles expressed in radian. And using angles in radian, we just use numbers as angles.

So for instance:

0° is 0 radian, which is put this way, too: 0 rad, for short. And we just 0 as 0 rad.

180° is π rad, where π is the circular ratio, which is 3.141592... And we just use π.

360° is 2π rad. And we just use 2π.

And the ray can make as many turns as necessary either counterclockwise or clockwise.

So we can put it this way: **-360°·n ≤ θ ≤ 360°·n ⇔ -2nπ ≤ θ ≤ 2nπ**, where *n* is an integer.

That is, we can get: -∞ < θ < ∞, where ∞ indicates infinity. So θ can be all real numbers.

And thus, we usually take values of *x* for numbers, too, which are however, angles.
So we want to keep in mind that the input values in a trig-function are actually angles.

Now, in sum, we have two categories in trigonometry.

One is static trigonometry.

In trigonometry static, we use normal right triangles applying or working with trig-ratios.

A normal right triangle has a side called the hypotenuse, together with two sides called the adjacent and the opposite perpendicular to each other. And of course, the lengths of all the sides are positive.

And in dynamic trigonometry, we mainly work with trig-functions.

In trigonometry dynamic, we use a transcendental right triangle, which keeps changing, and thus, is dynamic.

A right triangle transcendental, too, has a side called the hypotenuse, together with two sides called the adjacent and the opposite perpendicular to each other.

However, either or both of the adjacent and the opposite can be negative or even 0 as well as positive.

And in trig-functions, we have three basic ones as follows:

One is a sine function, where the expression part is **sin x**. Another is a cosine function, where the expression part is **cos x**. And the other is a tangent function, where the expression part is **tan x**. And in each, x usually takes angles in radian.

Usually though, we just take as numbers, angles in radian.

So for instance, assuming the domain is a set of all angles, and f is a sine function, we can put the trig-function f the way as follows: $y = f(x) = $ **sin** x for x real.

So what do we mean by 'x real'?

It means the input variable x takes all real numbers, i.e., any real number. And putting angles in radian, we express angles by means of numbers.

For instance:

$\pm 1° = \pm\pi/180, \quad \pm 30° = \pm\pi/6, \quad \pm 45° = \pm\pi/4, \quad \pm 60° = \pm\pi/3, \quad \pm 90° = \pm\pi/2, \quad \pm 180° = \pm\pi,$

etc. where $\pi = 3.141\ldots$

And in the function f, x takes in fact, an angle.

So 'x real' means that x can take any angle.

And thus, the domain of f is a set of all angles. What then, about the range?

The range is a set of all numbers from -1 to 1, and each of the numbers is a trig-ratio.

So the range of f can be put this way, too: **-1 $\leq y \leq$ 1**, or **$|y| \leq$ 1**.

Note however, if the domain is not a set of all real numbers, the range can be other than the one above.

And the same is true for a cosine function, too.

So for instance, assuming the domain is a set of all angles, and g is a cosine function, we can put the trig-function g the way as follows: $y = g(x) = \cos x$ for x real.

What then, about the range?

The range is a set of all numbers from -1 to 1, and all the number are trig-ratios.

So the range of g can be put this way, too: $-1 \leq y \leq 1$, or $|y| \leq 1$.

As in the case of the sine function f however, if the domain is not a set of all real numbers, it can be the case where the range is not the one above.

What then, about a tangent function?

For instance, assuming the domain is a set of all angles, and h is a tangent function, we can put the trig-function h the way as follows: $y = h(x) = \tan x$ for x real.

What then, about the range?

Unlike the sine and cosine functions above, the range of the tangent function h is a set of all real numbers, each of which is a trig-ratio, of course.

Note however, if the domain of a tangent function is not a set of all real numbers, it can be the case where the range is not a set of all real numbers.

Examples 1 in Trigonometry Dynamic

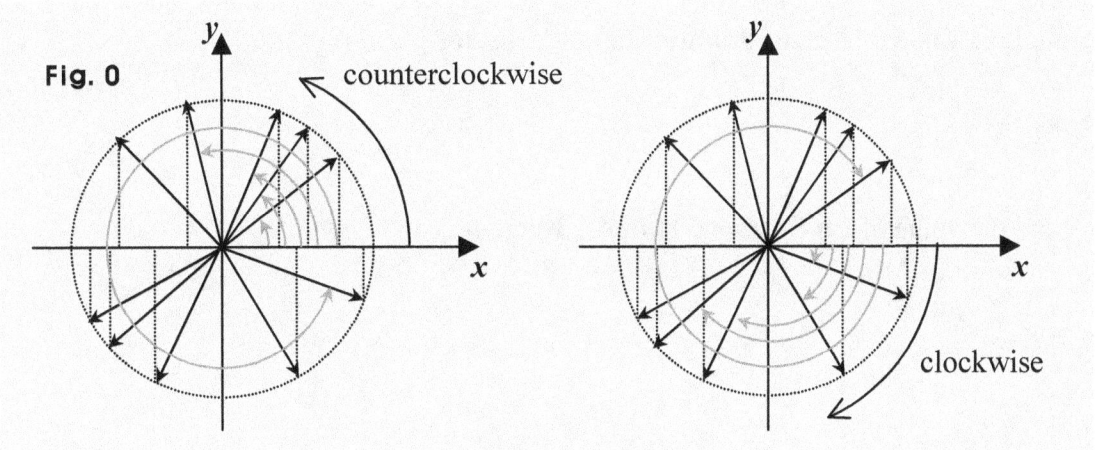

In trigonometry dynamic, the ray turning makes governing angles.

If turning clockwise, it makes angles negative, and if counterclockwise, it makes angles positive. And of course, if no turning, the angle made is 0.

So a governing angle can be 0° or any angle positive or negative.

In each right triangle, the ray is of length 1, and is the hypotenuse.

And (x, y) is the terminal point, so x is the adjacent, and y is the opposite.

And thus, assuming θ is a governing angle, we get: **sin** $\theta = y$, **cos** $\theta = x$, and **tan** $\theta = y/x$.

So if the ray is in the first quadrant, all the three trig-ratios > 0, since x and y both > 0.

In the second, only the sines > 0, since $x < 0$, and $y > 0$.

In the third, only the tangents > 0, since $x < 0$, and $y < 0$.

And in the fourth, only the cosines > 0, since $x > 0$, and $y < 0$.

0. Assuming $180° < \theta < 270°$, simplify the expression below:

$$\sqrt{\sin^2\theta} + \sqrt[3]{(\sin\theta+\cos\theta)^3} - \sqrt[4]{(\cos\theta+\tan\theta+1)^4}$$

1. Find the value of each as follows: **sin 420°, cos 960°,** and **tan (-1200°).**

2. Find the value of **sin** $\{n\pi/2 + (-1)^n(\pi/6)\}$ where **n** is an integer.

Suggestions or Solutions
To the **Problem** in the Example **0**

Assuming $180° < \theta < 270°$, simplify the expression below:

$$\sqrt{\sin^2\theta} + \sqrt[3]{(\sin\theta + \cos\theta)^3} - \sqrt[4]{(\cos\theta + \tan\theta + 1)^4}$$

We know, for instance, $\sqrt{2^2} = 2$.

So assuming $P = \sqrt{\sin^2\theta} + \sqrt[3]{(\sin\theta + \cos\theta)^3} - \sqrt[4]{(\cos\theta + \tan\theta + 1)^4}$, can we just P this way, too: $P = \sin\theta + (\sin\theta + \cos\theta) - (\cos\theta + \tan\theta + 1) = 2\sin\theta - \tan\theta - 1$?

We can do so if for instance, we have: $0° < \theta < 90°$.

If however, for instance, we have: $200° < \theta < 220°$, we cannot just remove the radical signs (the root signs) the way above.

That is to say that, for instance, we cannot just set: $\sqrt{\sin^2\theta} = \sin\theta$.   Why not?

That's because for instance, $\sqrt{x^2}$ is not just x but $|x|$.

It's because x can be negative, and we have: $\sqrt{x^2} \geq 0$.

And if $180° < \theta < 270°$, we get: $-1 < \sin\theta < 0$.

So if $180° < \theta < 270°$, we get: $\sqrt{\sin^2\theta} = -\sin\theta$, since: $0 < \sqrt{\sin^2\theta} < 1$.

And in fact, we want to set: $\sqrt[n]{x^n} = |x|$ if n is an integer even. It's because if n is an even integer, we get: $\sqrt[n]{x^n} \geq 0$, and x can be negative as well as 0 or positive. That is, if n is even, the sign of $\sqrt[n]{x^n}$ can be other than that of x. So for instance, $\sqrt[4]{(-2)^4} = 2$, and not -2.

If however, n is odd, we can just simply set: $\sqrt[n]{x^n} = x$. It's because the sign of $\sqrt[n]{x^n}$ is the same as that of x. For instance, $\sqrt[3]{(-3)^3} = -3$. So we want to be careful when removing radical signs.

If not quite sure of radicals, refer to **ALGEBRA EXAMPLES POWERS AND LOGARITHMS**.

Now, if $180° < \theta < 270°$, we get: $-1 < \sin \theta < 0$, $-1 < \cos \theta < 0$, and $\tan \theta > 0$. How come though?

That's because if $180° < \theta < 270°$, the terminal ray turning in the *x-y* plane is in the third quadrant, the *x*-coordinate of the terminal point is negative, and so is the *y*-coordinate. So what?

Assuming the ray is of length 1, and θ is the angle made by the ray, we can put the basic trig-ratios the way as follows: $\sin \theta$ is the *y*-coordinate of the terminal point, $\cos \theta$ is the *x*-coordinate of the terminal point, and $\tan \theta$ is the *y*-coordinate over the *x*-coordinate.

In short, if (x, y) is the terminal point of the ray, $\sin \theta = y$, $\cos \theta = x$, and $\tan \theta = y/x$.

And if the ray is in the third quadrant, that is, if $180° < \theta < 270°$, we get: $x < 0$, and $y < 0$.

And we know that:

If $\theta = 180°$, that is, π, we get: $y = 0$, and $x = -1$. So $\sin \theta = 0$, $\cos \theta = -1$, and $\tan \theta = 0$ since $\tan \theta = y/x$ and $y = 0$..

If $\theta = 270°$, that is, $3\pi/2$, we get: $y = -1$, and $x = 0$. So $\sin \theta = -1$, $\cos \theta = 0$, and $\tan \theta$ is not defined since $\tan \theta = y/x$ and $x = 0$.

And putting in graphs, the curves of $\sin \theta$, $\cos \theta$, and $\tan \theta$, we can see better how it is the case where if $\pi < \theta < 3\pi/2$, we get: $-1 < \sin \theta < 0$, $-1 < \cos \theta < 0$, and $\tan \theta > 0$.

We will cover how to construct those curves in the sections beginning the next section. And the curves will look like the ones below:

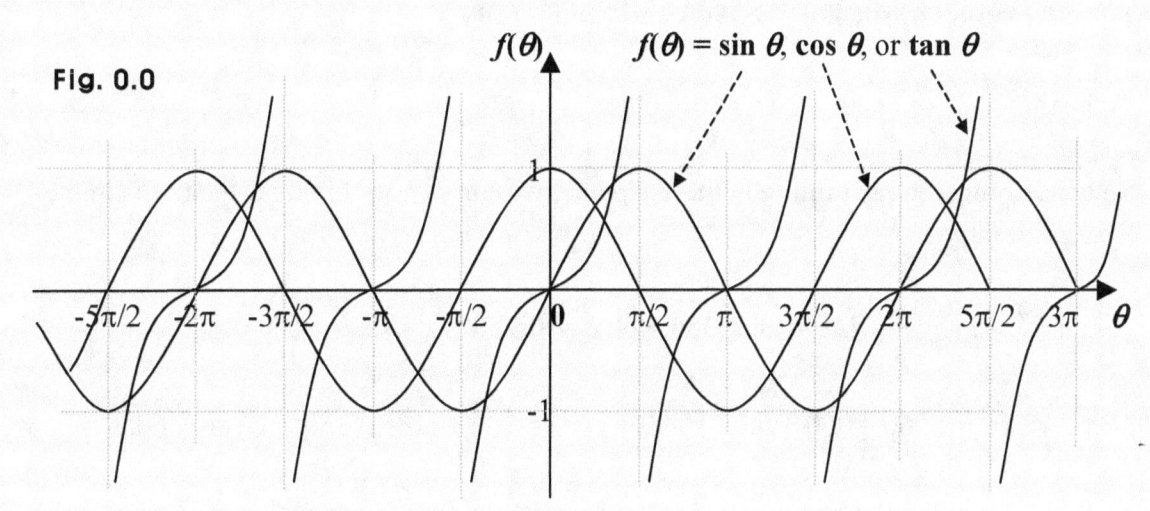

Now, we have: $P = \sqrt{\sin^2\theta} + \sqrt[3]{(\sin\theta + \cos\theta)^3} - \sqrt[4]{(\cos\theta + \tan\theta + 1)^4}$.

And we have: **-1 < sin θ < 0, -1 < cos θ < 0, and tan θ > 0**, since $\pi < \theta < 3\pi/2$.

So we get: **cos θ + 1 > 0** since **-1 < cos θ < 0**, and **cos θ + tan θ + 1 > 0** since **tan θ > 0**.

Thus, we get: **P = -sin θ + (sin θ + cos θ) – (cos θ + tan θ + 1) = -tan θ – 1**.

So we get: **P = -tan θ – 1 if 180° < θ < 270°**.

Suggestions or Solutions
To the **Problem** in the Example **1**

Find the values of sin 420°, cos 960°, and tan (-1200°).

To begin with, we know: 420 = 360 + 60.

So considering the ray turning in the *x-y* plane, we can say, in this case, the ray is in the first quadrant.

And thus, we get: **sin 420° = sin (360 + 60)° = sin 60° = $\frac{\sqrt{3}}{2}$.**

Next, we know: 960 = 900 + 60 = 10·90 + 60 = 5·180 + 60 = 2·360 + 1·180 + 60.

So the ray is in the third quadrant. And putting the ray in the *x-y* plane, we get:

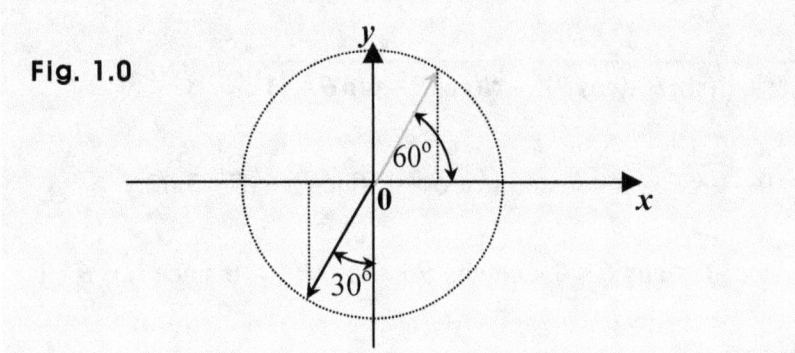

Fig. 1.0

So if the ray is of length 1, and **(*x*, *y*)** is the terminal point, we get: **cos (180 + 60)° = *x*,** and **cos 60° = -*x*.** How come?

That's because the cosine is the *x*-coordinate at the terminal point.

And thus, **cos 960° = cos (180 + 60)° = -cos 60° = -1/2,** since **cos 60° = 1/2.**

And next, we know: 1200 = 13·90 + 30 = 6·180 + 90 + 30 = 3·360 + 90 + 30. So?

So -1200 = -3·360 – 90 – 30, and thus, the ray is in the third quadrant.

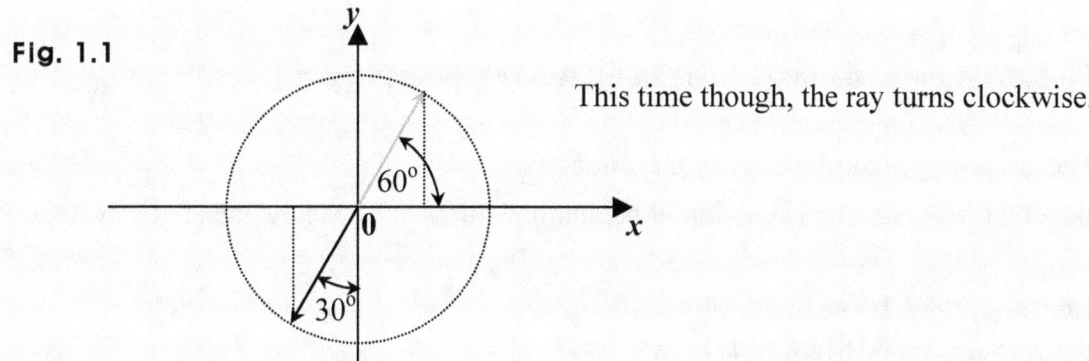

Fig. 1.1

This time though, the ray turns clockwise.

So assuming **tan (-90 – 30)° = y/x**, we get: **tan 60° = y/x**, too. How come?

That's because the tangent is: the opposite over the adjacent, and thus, is: the *y*-coordinate over the *x*-coordinate, at the terminal point, of course.

And thus, **tan (-1200°) = tan (-90 – 30)° = tan 60° = √3**.

Suggestions or Solutions
To the **Problem** in the Example **2**

Find the value of sin $\{n\pi/2 + (-1)^n(\pi/6)\}$ where n is an integer.

The governing angle depends on the value of **n**.

So let's first, put some values into **n** beginning with 0, and see how the angle changes.

- $n = 0 \Rightarrow n\pi/2 + (-1)^n(\pi/6) = \pi/6$.
- $n = 1 \Rightarrow n\pi/2 + (-1)^n(\pi/6) = \pi/2 - \pi/6 = \pi/3$.
- $n = 2 \Rightarrow n\pi/2 + (-1)^n(\pi/6) = 2\pi/2 + \pi/6 = \pi + \pi/6$.
- $n = 3 \Rightarrow n\pi/2 + (-1)^n(\pi/6) = 3\pi/2 - \pi/6 = \pi + \pi/2 - \pi/6 = \pi + \pi/3$.
- $n = 4 \Rightarrow n\pi/2 + (-1)^n(\pi/6) = 4\pi/2 + \pi/6 = 2\pi + \pi/6$.
- $n = 5 \Rightarrow n\pi/2 + (-1)^n(\pi/6) = 5\pi/2 - \pi/6 = 2\pi + \pi/2 - \pi/6 = 2\pi + \pi/3$.
- $n = 6 \Rightarrow n\pi/2 + (-1)^n(\pi/6) = 6\pi/2 + \pi/6 = 2\pi + \pi + \pi/6$.
- $n = 7 \Rightarrow n\pi/2 + (-1)^n(\pi/6) = 7\pi/2 - \pi/6 = 2\pi + 3\pi/2 - \pi/6 = 2\pi + \pi + \pi/3$.
- $n = 8 \Rightarrow n\pi/2 + (-1)^n(\pi/6) = 8\pi/2 + \pi/6 = 4\pi + \pi/6$.
- $n = 9 \Rightarrow n\pi/2 + (-1)^n(\pi/6) = 9\pi/2 - \pi/6 = 4\pi + \pi/2 - \pi/6 = 4\pi + \pi/3$.
- $n = 10 \Rightarrow n\pi/2 + (-1)^n(\pi/6) = 10\pi/2 + \pi/6 = 4\pi + \pi + \pi/6$.
- $n = 11 \Rightarrow n\pi/2 + (-1)^n(\pi/6) = 11\pi/2 - \pi/6 = 4\pi + 3\pi/2 - \pi/6 = 4\pi + \pi + \pi/3$.
- $n = 12 \Rightarrow n\pi/2 + (-1)^n(\pi/6) = 12\pi/2 + \pi/6 = 6\pi + \pi/6$.

Now, taking a closer look at the sequence of angles above, we can notice that if **$n = 4k$** for **k** integer ≥ 0, the angles are no different from each other.

Fig. 2.0

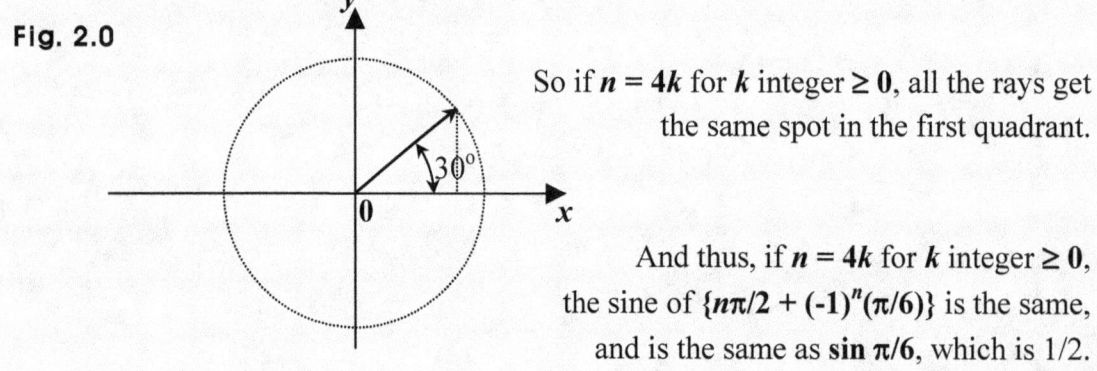

So if **$n = 4k$** for **k** integer ≥ 0, all the rays get the same spot in the first quadrant.

And thus, if **$n = 4k$** for **k** integer ≥ 0, the sine of $\{n\pi/2 + (-1)^n(\pi/6)\}$ is the same, and is the same as **sin $\pi/6$**, which is 1/2.

So if **$n = 4k$** for **k** integer ≥ 0, sin $\{n\pi/2 + (-1)^n(\pi/6)\} = $ sin $\pi/6 = 1/2$.

Next, we can notice that if $n = 4k + 1$ for k integer ≥ 0, the angles are again, no different from each other.

$n = 1 \Rightarrow n\pi/2 + (-1)^n(\pi/6) = \pi/2 - \pi/6 = \pi/3.$

$n = 5 \Rightarrow n\pi/2 + (-1)^n(\pi/6) = 5\pi/2 - \pi/6 = 2\pi + \pi/2 - \pi/6 = 2\pi + \pi/3.$

$n = 9 \Rightarrow n\pi/2 + (-1)^n(\pi/6) = 9\pi/2 - \pi/6 = 4\pi + \pi/2 - \pi/6 = 4\pi + \pi/3.$

Fig. 2.1

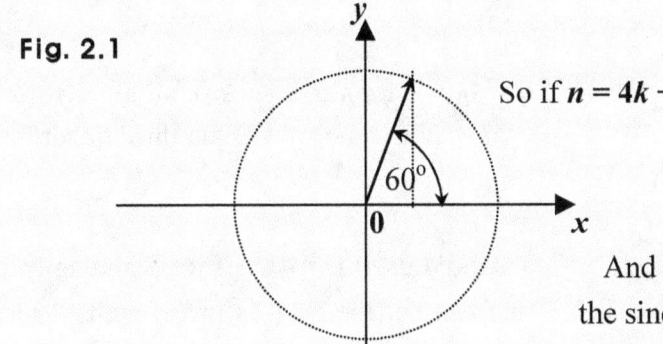

So if $n = 4k + 1$ for k integer ≥ 0, all the rays get the same spot in the first quadrant.

And thus, if $n = 4k + 1$ for k integer ≥ 0, the sine of $\{n\pi/2 + (-1)^n(\pi/6)\}$ is the same, and is the same as $\sin (\pi/3)$, which is $\frac{\sqrt{3}}{2}$.

So if $n = 4k + 1$ for k integer ≥ 0, $\sin \{n\pi/2 + (-1)^n(\pi/6)\} = \sin \pi/3 = \frac{\sqrt{3}}{2}$.

Next, we can notice that if $n = 4k + 2$ for k integer ≥ 0, the angles are again, no different from each other.

$n = 2 \Rightarrow n\pi/2 + (-1)^n(\pi/6) = 2\pi/2 + \pi/6 = \pi + \pi/6.$

$n = 6 \Rightarrow n\pi/2 + (-1)^n(\pi/6) = 6\pi/2 + \pi/6 = 2\pi + \pi + \pi/6.$

$n = 10 \Rightarrow n\pi/2 + (-1)^n(\pi/6) = 10\pi/2 + \pi/6 = 4\pi + \pi + \pi/6.$

Fig. 2.2

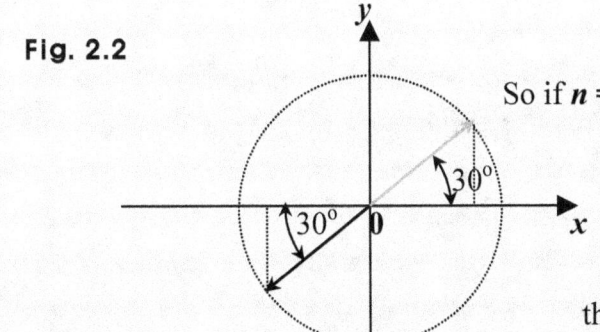

So if $n = 4k + 2$ for k integer ≥ 0, all the rays get the same spot in the third quadrant.

And thus, if $n = 4k + 2$ for k integer ≥ 0, the sine of $\{n\pi/2 + (-1)^n(\pi/6)\}$ is the same, and is the same as $\sin (\pi + \pi/6)$, which is -1/2. That's because $\sin (\pi + \pi/6) = -\sin \pi/6 = -1/2$.

So if $n = 4k + 2$ for k integer ≥ 0, $\sin \{n\pi/2 + (-1)^n(\pi/6)\} = -\sin \pi/6 = -1/2$.

And next, we can notice that if $n = 4k + 3$ for k integer ≥ 0, the angles are again, no different from each other.

$n = 3 \Rightarrow n\pi/2 + (-1)^n(\pi/6) = 3\pi/2 - \pi/6 = \pi + \pi/2 - \pi/6 = \pi + \pi/3$.

$n = 7 \Rightarrow n\pi/2 + (-1)^n(\pi/6) = 7\pi/2 - \pi/6 = 2\pi + 3\pi/2 - \pi/6 = 2\pi + \pi + \pi/3$.

$n = 11 \Rightarrow n\pi/2 + (-1)^n(\pi/6) = 11\pi/2 - \pi/6 = 4\pi + 3\pi/2 - \pi/6 = 4\pi + \pi + \pi/3$.

Fig. 2.3

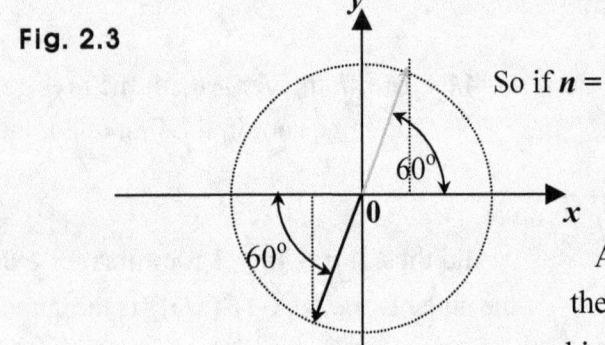

So if $n = 4k + 3$ for k integer ≥ 0, all the rays get the same spot in the third quadrant.

And thus, if $n = 4k + 3$ for k integer ≥ 0, the sine of $\{n\pi/2 + (-1)^n(\pi/6)\}$ is the same, and is the same as $\sin (\pi + \pi/3)$, which is $-\frac{\sqrt{3}}{2}$. That's because $\sin (\pi + \pi/3) = -\sin \pi/3 = -\frac{\sqrt{3}}{2}$.

So if $n = 4k + 3$ for k integer ≥ 0, $\sin \{n\pi/2 + (-1)^n(\pi/6)\} = -\sin \pi/3 = -\frac{\sqrt{3}}{2}$.

What if though, k is an integer negative?

Putting some negative values into n beginning with 0, we get a sequence below:

• $n = 0 \Rightarrow n\pi/2 + (-1)^n(\pi/6) = \pi/6$.

$n = -1 \Rightarrow n\pi/2 + (-1)^n(\pi/6) = -\pi/2 - \pi/6 = -2\pi/3$.

$n = -2 \Rightarrow n\pi/2 + (-1)^n(\pi/6) = -2\pi/2 + \pi/6 = -\pi + \pi/6$.

$n = -3 \Rightarrow n\pi/2 + (-1)^n(\pi/6) = -3\pi/2 - \pi/6 = -\pi - \pi/2 - \pi/6 = -\pi - 2\pi/3$.

• $n = -4 \Rightarrow n\pi/2 + (-1)^n(\pi/6) = -4\pi/2 + \pi/6 = -2\pi + \pi/6$.

$n = -5 \Rightarrow n\pi/2 + (-1)^n(\pi/6) = -5\pi/2 - \pi/6 = -2\pi - \pi/2 - \pi/6 = -2\pi - 2\pi/3$.

$n = -6 \Rightarrow n\pi/2 + (-1)^n(\pi/6) = -6\pi/2 + \pi/6 = -2\pi - \pi + \pi/6$.

$n = -7 \Rightarrow n\pi/2 + (-1)^n(\pi/6) = -7\pi/2 - \pi/6 = -2\pi - 3\pi/2 - \pi/6 = -2\pi - \pi - 2\pi/3$.

• $n = -8 \Rightarrow n\pi/2 + (-1)^n(\pi/6) = -8\pi/2 + \pi/6 = -4\pi + \pi/6$.

$n = -9 \Rightarrow n\pi/2 + (-1)^n(\pi/6) = -9\pi/2 - \pi/6 = -4\pi - \pi/2 - \pi/6 = -4\pi - 2\pi/3$.

$n = -10 \Rightarrow n\pi/2 + (-1)^n(\pi/6) = -10\pi/2 + \pi/6 = -4\pi - \pi + \pi/6$.

$n = -11 \Rightarrow n\pi/2 + (-1)^n(\pi/6) = -11\pi/2 - \pi/6 = -4\pi + 3\pi/2 - \pi/6 = -4\pi - \pi - 2\pi/3$.

• $n = -12 \Rightarrow n\pi/2 + (-1)^n(\pi/6) = -12\pi/2 + \pi/6 = -6\pi + \pi/6$.

Now, taking a closer look at the sequence of angles above, we can notice again, if $n = 4k$ for k integer ≤ 0, the angles are no different from each other.

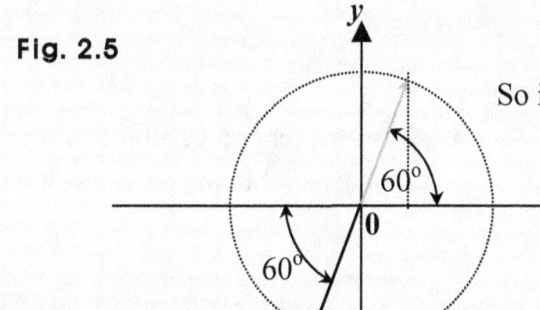

Fig. 2.4

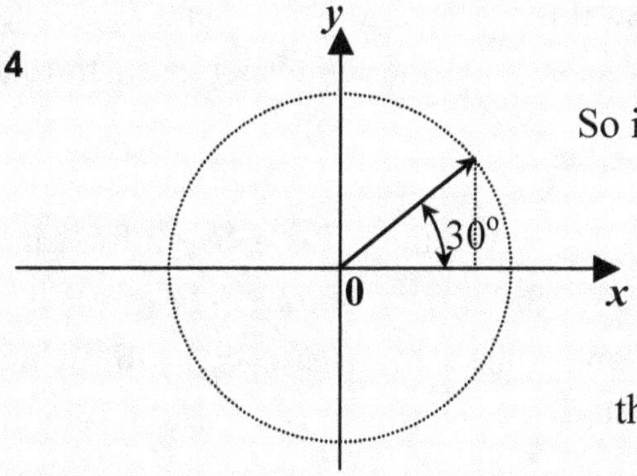

So if $n = 4k$ for k integer ≤ 0, all the rays get the same spot in the first quadrant.

And thus, if $n = 4k$ for k integer ≥ 0, the sine of $\{n\pi/2 + (-1)^n(\pi/6)\}$ is the same, and is the same as $\sin \pi/6$, which is 1/2.

So if $n = 4k$ for k integer ≤ 0, $\sin \{n\pi/2 + (-1)^n(\pi/6)\} = 1/2$.

Next, we can notice that if $n = 4k - 1$ for k integer ≤ 0, the angles are again, no different from each other.

$n = -1 \Rightarrow n\pi/2 + (-1)^n(\pi/6) = -\pi/2 - \pi/6 = -2\pi/3$.

$n = -5 \Rightarrow n\pi/2 + (-1)^n(\pi/6) = -5\pi/2 - \pi/6 = -2\pi - \pi/2 - \pi/6 = -2\pi - 2\pi/3$.

$n = -9 \Rightarrow n\pi/2 + (-1)^n(\pi/6) = -9\pi/2 - \pi/6 = -4\pi - \pi/2 - \pi/6 = -4\pi - 2\pi/3$.

Fig. 2.5

So if $n = 4k - 1$ for k integer ≤ 0, all the rays get the same spot in the third quadrant.

And thus, if $n = 4k - 1$ for k integer ≤ 0, the sine of $\{n\pi/2 + (-1)^n(\pi/6)\}$ is the same, and is the same as $-\sin (\pi/3)$, which is $-\frac{\sqrt{3}}{2}$.

So if $n = 4k - 1$ for k integer ≤ 0, $\sin \{n\pi/2 + (-1)^n(\pi/6)\} = -\sin \pi/3 = -\frac{\sqrt{3}}{2}$.

Next, we can notice that if $n = 4k - 2$ for k integer ≤ 0, the angles are again, no different from each other.

$n = -2 \Rightarrow n\pi/2 + (-1)^n(\pi/6) = -2\pi/2 + \pi/6 = -\pi + \pi/6$.

$n = -6 \Rightarrow n\pi/2 + (-1)^n(\pi/6) = -6\pi/2 + \pi/6 = -2\pi - \pi + \pi/6$.

$n = -10 \Rightarrow n\pi/2 + (-1)^n(\pi/6) = -10\pi/2 + \pi/6 = -4\pi - \pi + \pi/6$.

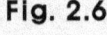

Fig. 2.6

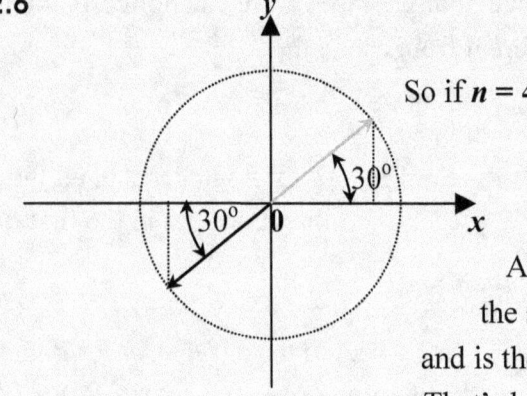

So if $n = 4k - 2$ for k integer ≤ 0, all the rays get the same spot in the third quadrant.

And thus, if $n = 4k - 2$ for k integer ≤ 0, the sine of $\{n\pi/2 + (-1)^n(\pi/6)\}$ is the same, and is the same as $\sin(-\pi + \pi/6)$, which is -1/2. That's because $\sin(-\pi + \pi/6) = -\sin\pi/6 = -1/2$.

So if $n = 4k - 2$ for k integer ≤ 0, $\sin\{n\pi/2 + (-1)^n(\pi/6)\} = -\sin\pi/6 = -1/2$.

And next, we can notice that if $n = 4k - 3$ for k integer ≤ 0, the angles are again, no different from each other.

$n = -3 \Rightarrow n\pi/2 + (-1)^n(\pi/6) = -3\pi/2 - \pi/6 = -\pi - \pi/2 - \pi/6 = -\pi - 2\pi/3$.

$n = -7 \Rightarrow n\pi/2 + (-1)^n(\pi/6) = -7\pi/2 - \pi/6 = -2\pi - 3\pi/2 - \pi/6 = -2\pi - \pi - 2\pi/3$.

$n = -11 \Rightarrow n\pi/2 + (-1)^n(\pi/6) = -11\pi/2 - \pi/6 = -4\pi + 3\pi/2 - \pi/6 = -4\pi - \pi - 2\pi/3$.

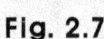

Fig. 2.7

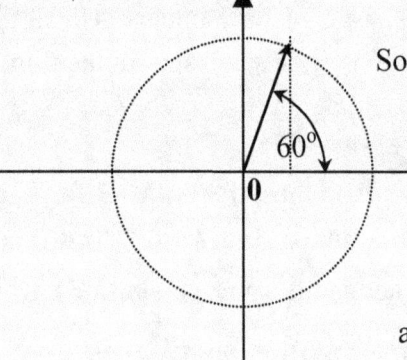

So if $n = 4k - 3$ for k integer ≤ 0, all the rays get the same spot in the third quadrant.

And thus, if $n = 4k - 3$ for k integer ≤ 0, the sine of $\{n\pi/2 + (-1)^n(\pi/6)\}$ is the same, and is the same as $\sin(-\pi - 2\pi/3)$, which is $\frac{\sqrt{3}}{2}$. That's because $\sin(-\pi - 2\pi/3) = \sin\pi/3 = \frac{\sqrt{3}}{2}$.

Thus, if $n = 4k - 3$ for k integer ≤ 0, $\sin\{n\pi/2 + (-1)^n(\pi/6)\} = \sin\pi/3 = \frac{\sqrt{3}}{2}$.

So putting together now, the case where k is greater than or equal to 0, and the case where k is less than 0, we can put them the way below:

If $n = 4k$ for k integer, $\sin\{n\pi/2 + (-1)^n(\pi/6)\} = 1/2$.

If $n = 4k + 1$ for k integer ≥ 0, $\sin\{n\pi/2 + (-1)^n(\pi/6)\} = \frac{\sqrt{3}}{2}$.

If $n = 4k - 3$ for k integer ≤ 0, $\sin\{n\pi/2 + (-1)^n(\pi/6)\} = \frac{\sqrt{3}}{2}$.

If $n = 4k + 3$ for k integer ≥ 0, $\sin\{n\pi/2 + (-1)^n(\pi/6)\} = -\frac{\sqrt{3}}{2}$.

If $n = 4k - 1$ for k integer ≤ 0, $\sin\{n\pi/2 + (-1)^n(\pi/6)\} = -\frac{\sqrt{3}}{2}$.

If $n = 4k + 2$ for k integer ≥ 0, $\sin\{n\pi/2 + (-1)^n(\pi/6)\} = -1/2$.

And if $n = 4k - 2$ for k integer ≤ 0, $\sin\{n\pi/2 + (-1)^n(\pi/6)\} = -1/2$.

Taking a closer look at however, for instance, the case where $n = 4k + 1$ for k integer ≥ 0, and the case where $n = 4k - 3$ for k integer ≤ 0, we can just put both together the way as follows: $n = 4k + 1$ for k integer. How come?

If $n = 4k + 1$ for k integer ≥ 0, we get: $n = 1, 5, 9, 13, 17$, and so on.

If $n = 4k - 3$ for k integer ≤ 0, we get: $n = -3, -7, -11, -15$, and so forth.

And if $n = 4k + 1$ for k integer, that is, $k = 0, 1, -1, 2, -2, 3, -3, 4, -4$, etc., we get: $n = 1, 5, -3, 9, -7, 13, -11, 17, -15, \ldots$

And the same is true for the other cases where we get the same ratios.
So we can put together all the cases above the way below:

If $n = 4k$ for k integer, $\sin\{n\pi/2 + (-1)^n(\pi/6)\} = 1/2$.

If $n = 4k + 1$ for k integer, $\sin\{n\pi/2 + (-1)^n(\pi/6)\} = \frac{\sqrt{3}}{2}$.

If $n = 4k + 2$ for k integer, $\sin\{n\pi/2 + (-1)^n(\pi/6)\} = -1/2$.

And if $n = 4k + 3$ for k integer, $\sin\{n\pi/2 + (-1)^n(\pi/6)\} = -\frac{\sqrt{3}}{2}$.

And in fact, we can put together all the cases above the way below, too:

If $n = 4k$ for k integer, $\sin\{n\pi/2 + (-1)^n(\pi/6)\} = 1/2$.

If $n = 4k - 1$ for k integer, $\sin\{n\pi/2 + (-1)^n(\pi/6)\} = -\frac{\sqrt{3}}{2}$.

if $n = 4k - 2$ for k integer, $\sin\{n\pi/2 + (-1)^n(\pi/6)\} = -1/2$.

And if $n = 4k - 3$ for k integer, $\sin\{n\pi/2 + (-1)^n(\pi/6)\} = \frac{\sqrt{3}}{2}$.

How come?

If $n = 4k + 1$ for k integer ≥ 0, we get: $n = 1, 5, 9, 13$, and so on.

If $n = 4k - 3$ for k integer ≤ 0, we get: $n = -3, -7, -11, -15, -19$, and so forth.

And if $n = 4k - 3$ for k integer, that is, $k = 0, 1, -1, 2, -2, 3, -3, 4, -4$, etc., we get: $n = -3, 1, -7, 5, -11, 9, -15, 13, -19, \ldots$

And the same is true for the other cases where we get the same ratios.

So putting the threads together, we want to consider the four cases where $n = 4k$, $4k + 1$, $4k + 2$, and $4k + 3$ for k is an integer.

Examples 2 in Trigonometry Dynamic

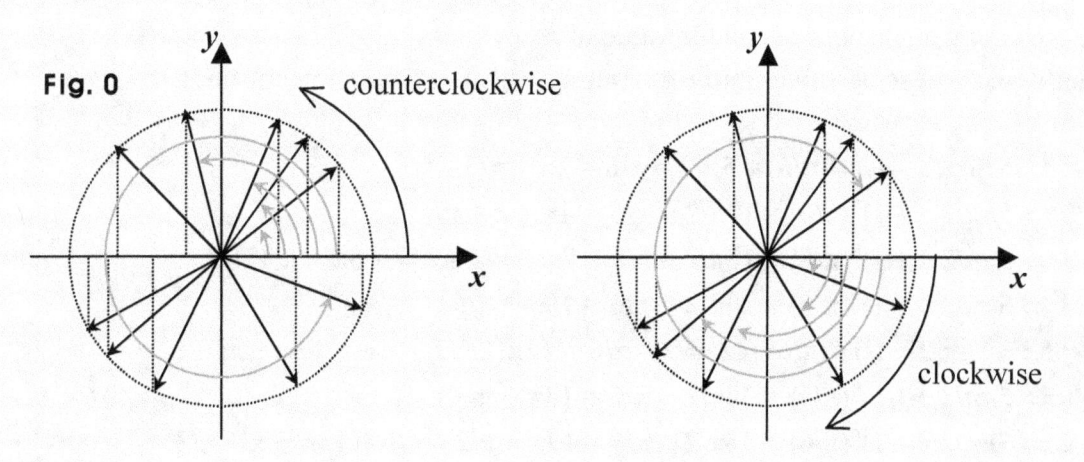

Fig. 0

In trigonometry dynamic, the ray turning makes governing angles.

If turning clockwise, it makes angles negative, and if counterclockwise, it makes angles positive. And of course, if no turning, the angle made is 0.

So a governing angle can be $0°$ or any angle positive or negative.

In each right triangle, the ray is of length 1, and is the hypotenuse.

And (x, y) is the terminal point, so x is the adjacent, and y is the opposite.

And thus, assuming θ is a governing angle, we get: $\sin\theta = y$, $\cos\theta = x$, and $\tan\theta = y/x$.

So if the ray is in the first quadrant, all the three trig-ratios > 0, since x and y both > 0.

In the second, only the sines > 0, since $x < 0$, and $y > 0$.

In the third, only the tangents > 0, since $x < 0$, and $y < 0$.

And in the fourth, only the cosines > 0, since $x > 0$, and $y < 0$.

0.   Find the value of $\sin\{5n\pi/2 + (-1)^{n}(\pi/6)\}$ where n is a positive integer.

1.   Find the value of $\sin\{5n\pi + (-1)^{n}(\pi/6)\}$ where n is a nonnegative integer.

Suggestions or Solutions
To the **Problem** in the Example **0**

Find the value of sin $\{5n\pi/2 + (-1)^n(\pi/6)\}$ where *n* is a positive integer.

The governing angle depends on the value of **n**.
So let's first, put some values into **n** beginning with 1, and see how the angle changes.

- $n = 1 \Rightarrow 5n\pi/2 + (-1)^n(\pi/6) = 5\pi/2 - \pi/6 = 2\pi + \pi/3.$
- $n = 2 \Rightarrow 5n\pi/2 + (-1)^n(\pi/6) = 10\pi/2 + \pi/6 = 4\pi + \pi + \pi/6.$
- $n = 3 \Rightarrow 5n\pi/2 + (-1)^n(\pi/6) = 15\pi/2 - \pi/6 = 7\pi + \pi/2 - \pi/6 = 6\pi + \pi + \pi/3.$
- $n = 4 \Rightarrow 5n\pi/2 + (-1)^n(\pi/6) = 20\pi/2 + \pi/6 = 10\pi + \pi/6.$
- $n = 5 \Rightarrow 5n\pi/2 + (-1)^n(\pi/6) = 25\pi/2 - \pi/6 = 12\pi + \pi/2 - \pi/6 = 12\pi + \pi/3.$
- $n = 6 \Rightarrow 5n\pi/2 + (-1)^n(\pi/6) = 30\pi/2 + \pi/6 = 14\pi + \pi + \pi/6.$
- $n = 7 \Rightarrow 5n\pi/2 + (-1)^n(\pi/6) = 35\pi/2 - \pi/6 = 17\pi + \pi/2 - \pi/6 = 16\pi + \pi + \pi/3.$
- $n = 8 \Rightarrow 5n\pi/2 + (-1)^n(\pi/6) = 40\pi/2 + \pi/6 = 20\pi + \pi/6.$
- $n = 9 \Rightarrow 5n\pi/2 + (-1)^n(\pi/6) = 45\pi/2 - \pi/6 = 22\pi + \pi/2 - \pi/6 = 22\pi + \pi/3.$
- $n = 10 \Rightarrow 5n\pi/2 + (-1)^n(\pi/6) = 50\pi/2 + \pi/6 = 24\pi + \pi + \pi/6.$
- $n = 11 \Rightarrow 5n\pi/2 + (-1)^n(\pi/6) = 55\pi/2 - \pi/6 = 26\pi + \pi + \pi/2 - \pi/6 = 26\pi + \pi + \pi/3.$
- $n = 12 \Rightarrow 5n\pi/2 + (-1)^n(\pi/6) = 60\pi/2 + \pi/6 = 30\pi + \pi/6.$
- $n = 13 \Rightarrow 5n\pi/2 + (-1)^n(\pi/6) = 65\pi/2 - \pi/6 = 32\pi + \pi/3.$

Now, taking a closer look at the sequence above, we can notice that if $n = 4k + 1$ for **k** integer ≥ 0, the angles are no different from each other.

Fig. 0.0

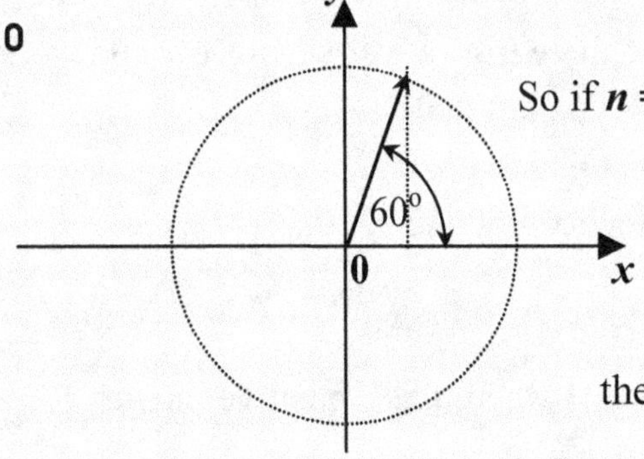

So if $n = 4k + 1$ for **k** integer ≥ 0, all the rays get the same spot in the first quadrant.

And thus, if $n = 4k + 1$ for **k** integer ≥ 0, the sine of $\{5n\pi/2 + (-1)^n(\pi/6)\}$ is the same, and is the same as **sin $\pi/3$**, which is $\frac{\sqrt{3}}{2}$.

So if $n = 4k + 1$ for **k** integer ≥ 0, sin $\{5n\pi/2 + (-1)^n(\pi/6)\} = \sin \pi/3 = \frac{\sqrt{3}}{2}$.

Next, we can notice that if $n = 4k + 2$ for k integer ≥ 0, the angles are again, no different from each other.

$n = 2 \Rightarrow 5n\pi/2 + (-1)^n(\pi/6) = 10\pi/2 + \pi/6 = 4\pi + \pi + \pi/6$.

$n = 6 \Rightarrow 5n\pi/2 + (-1)^n(\pi/6) = 30\pi/2 + \pi/6 = 14\pi + \pi + \pi/6$.

$n = 10 \Rightarrow 5n\pi/2 + (-1)^n(\pi/6) = 50\pi/2 + \pi/6 = 24\pi + \pi + \pi/6$.

Fig. 0.1

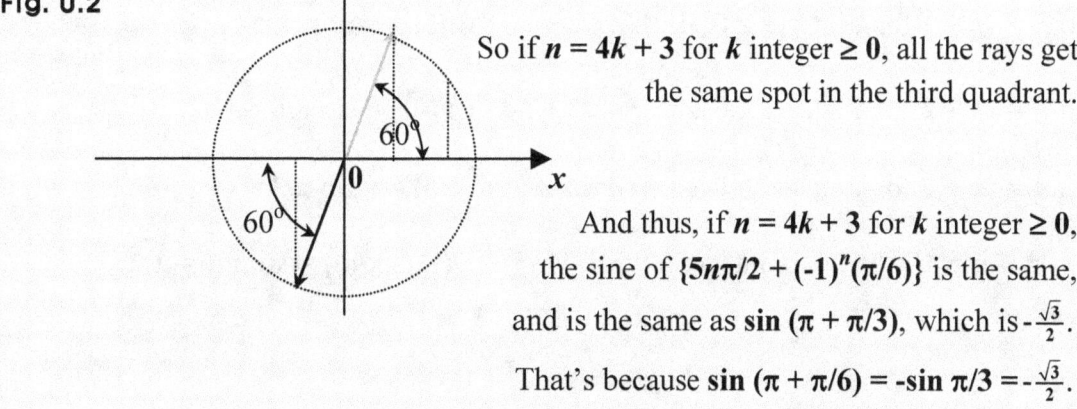

So if $n = 4k + 2$ for k integer ≥ 0, all the rays get the same spot in the third quadrant.

And thus, if $n = 4k + 2$ for k integer ≥ 0, the sine of $\{5n\pi/2 + (-1)^n(\pi/6)\}$ is the same, and is the same as **-sin $(\pi/6)$**, which is -1/2.

So if $n = 4k + 2$ for k integer ≥ 0, sin $\{5n\pi/2 + (-1)^n(\pi/6)\}$ = **-sin $\pi/6$** = **-1/2**.

Next, we can notice that if $n = 4k + 3$ for k integer ≥ 0, the angles are again, no different from each other.

$n = 3 \Rightarrow 5n\pi/2 + (-1)^n(\pi/6) = 15\pi/2 - \pi/6 = 7\pi + \pi/2 - \pi/6 = 6\pi + \pi + \pi/3$.

$n = 7 \Rightarrow 5n\pi/2 + (-1)^n(\pi/6) = 35\pi/2 - \pi/6 = 17\pi + \pi/2 - \pi/6 = 16\pi + \pi + \pi/3$.

$n = 11 \Rightarrow 5n\pi/2 + (-1)^n(\pi/6) = 55\pi/2 - \pi/6 = 26\pi + \pi + \pi/2 - \pi/6 = 26\pi + \pi + \pi/3$.

Fig. 0.2

So if $n = 4k + 3$ for k integer ≥ 0, all the rays get the same spot in the third quadrant.

And thus, if $n = 4k + 3$ for k integer ≥ 0, the sine of $\{5n\pi/2 + (-1)^n(\pi/6)\}$ is the same, and is the same as **sin $(\pi + \pi/3)$**, which is $-\frac{\sqrt{3}}{2}$.

That's because **sin $(\pi + \pi/6)$** = -sin $\pi/3$ = $-\frac{\sqrt{3}}{2}$.

So if $n = 4k + 3$ for k integer ≥ 0, sin $\{5n\pi/2 + (-1)^n(\pi/6)\}$ = -sin $\pi/3$ = $-\frac{\sqrt{3}}{2}$.

116

And next, we can notice that if $n = 4k + 4 = 4(k + 1)$ for k integer ≥ 0, the angles are again, no different from each other.

$n = 4 \Rightarrow 5n\pi/2 + (-1)^n(\pi/6) = 20\pi/2 + \pi/6 = 10\pi + \pi/6.$

$n = 8 \Rightarrow 5n\pi/2 + (-1)^n(\pi/6) = 40\pi/2 + \pi/6 = 20\pi + \pi/6.$

$n = 12 \Rightarrow 5n\pi/2 + (-1)^n(\pi/6) = 60\pi/2 + \pi/6 = 30\pi + \pi/6.$

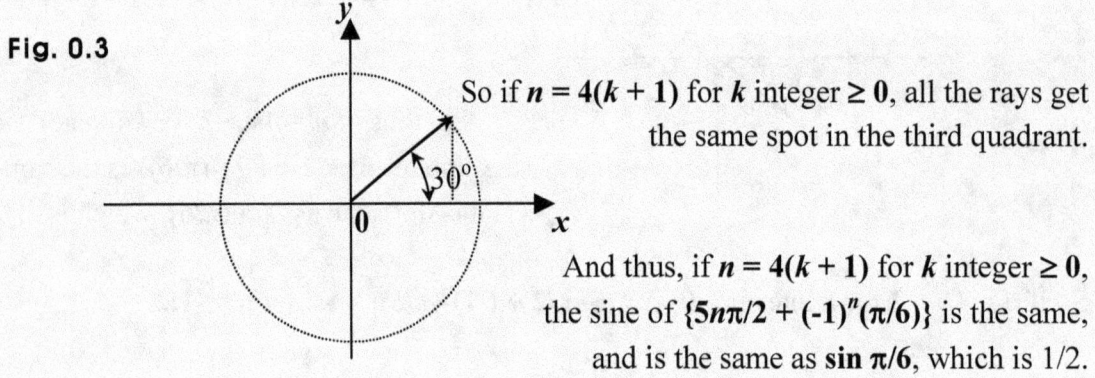

Fig. 0.3

So if $n = 4(k + 1)$ for k integer ≥ 0, all the rays get the same spot in the third quadrant.

And thus, if $n = 4(k + 1)$ for k integer ≥ 0, the sine of $\{5n\pi/2 + (-1)^n(\pi/6)\}$ is the same, and is the same as **sin $\pi/6$**, which is 1/2.

So if $n = 4(k + 1)$ for k integer ≥ 0, sin $\{5n\pi/2 + (-1)^n(\pi/6)\}$ = sin $\pi/6$ = 1/2.

So putting together now, all the four cases above, we can put them the way below:

If $n = 4k + 1$ for k integer ≥ 0, sin $\{5n\pi/2 + (-1)^n(\pi/6)\}$ = sin $\pi/3$ = $\frac{\sqrt{3}}{2}$.

If $n = 4k + 2$ for k integer ≥ 0, sin $\{5n\pi/2 + (-1)^n(\pi/6)\}$ = -sin $\pi/6$ = -1/2.

If $n = 4k + 3$ for k integer ≥ 0, sin $\{5n\pi/2 + (-1)^n(\pi/6)\}$ = -sin $\pi/3$ = -$\frac{\sqrt{3}}{2}$.

And if $n = 4(k + 1)$ for k integer ≥ 0, sin $\{5n\pi/2 + (-1)^n(\pi/6)\}$ = sin $\pi/6$ = 1/2.

Suggestions or Solutions
To the **Problem** in the Example **1**

Find the value of sin $\{5n\pi + (-1)^n(\pi/6)\}$ where n is a nonnegative integer.

The governing angle depends on the value of **n**.
So let's first, put some values into **n** beginning with 0, and see how the angle changes.

- $n = 0 \Rightarrow 5n\pi + (-1)^n(\pi/6) = \pi/6$.

  $n = 1 \Rightarrow 5n\pi + (-1)^n(\pi/6) = 5\pi - \pi/6 = 4\pi + 5\pi/6$.

- $n = 2 \Rightarrow 5n\pi + (-1)^n(\pi/6) = 10\pi + \pi/6$.

  $n = 3 \Rightarrow 5n\pi + (-1)^n(\pi/6) = 15\pi - \pi/6 = 14\pi + 5\pi/6$.

- $n = 4 \Rightarrow 5n\pi + (-1)^n(\pi/6) = 20\pi + \pi/6$.

  $n = 5 \Rightarrow 5n\pi + (-1)^n(\pi/6) = 25\pi - \pi/6 = 24\pi + 5\pi/6$.

- $n = 6 \Rightarrow 5n\pi + (-1)^n(\pi/6) = 30\pi + \pi/6$.

Now, we can notice that if $n = 2k$ for k integer ≥ 0, the angles are no different from each other.

Fig. 1.0

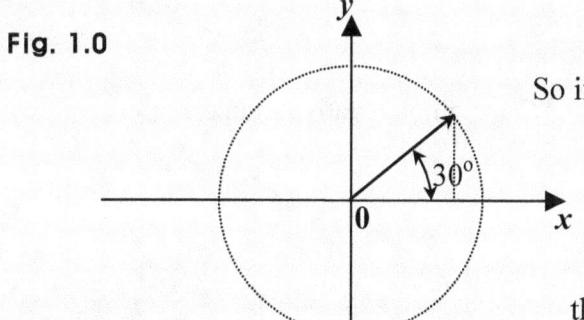

So if $n = 2k$ for k integer ≥ 0, all the rays get the same spot in the first quadrant.

And thus, if $n = 2k$ for k integer ≥ 0, the sine of $\{5n\pi + (-1)^n(\pi/6)\}$ is the same, and is the same as **sin $\pi/6$**, which is 1/2.

So if $n = 2k$ for k integer ≥ 0, sin $\{5n\pi + (-1)^n(\pi/6)\}$ = sin $\pi/6$ = 1/2.

Next, we can notice that if $n = 2k + 1$ for k integer ≥ 0, the angles are again, no different from each other.

$n = 1 \Rightarrow 5n\pi + (-1)^n(\pi/6) = 5\pi - \pi/6 = 4\pi + 5\pi/6$.

$n = 3 \Rightarrow 5n\pi + (-1)^n(\pi/6) = 15\pi - \pi/6 = 14\pi + 5\pi/6$.

$n = 5 \Rightarrow 5n\pi + (-1)^n(\pi/6) = 25\pi - \pi/6 = 24\pi + 5\pi/6$.

Fig. 1.1

So if $n = 2k + 1$ for k integer ≥ 0, all the rays get the same spot in the second quadrant.

And thus, if $n = 2k + 1$ for k integer ≥ 0, the sine of $\{5n\pi + (-1)^n(\pi/6)\}$ is the same, and is the same as $\sin(5\pi/6)$, which is 1/2. That's because $\sin(5\pi/6) = \sin \pi/6$

So if $n = 2k + 1$ for k integer ≥ 0, $\sin\{5n\pi + (-1)^n(\pi/6)\} = \sin 5\pi/6 = \sin \pi/6 = 1/2$.

So putting together now, the two cases above, we can put them the way below:

If $n = 2k$ for k integer ≥ 0, $\sin\{5n\pi + (-1)^n(\pi/6)\} = 1/2$.

And also, if $n = 2k + 1$ for k integer ≥ 0, $\sin\{5n\pi + (-1)^n(\pi/6)\} = 1/2$.

And thus, we get: $\sin\{5n\pi + (-1)^n(\pi/6)\} = 1/2$ for n nonnegative integer.

Examples 3 in Trigonometry Dynamic

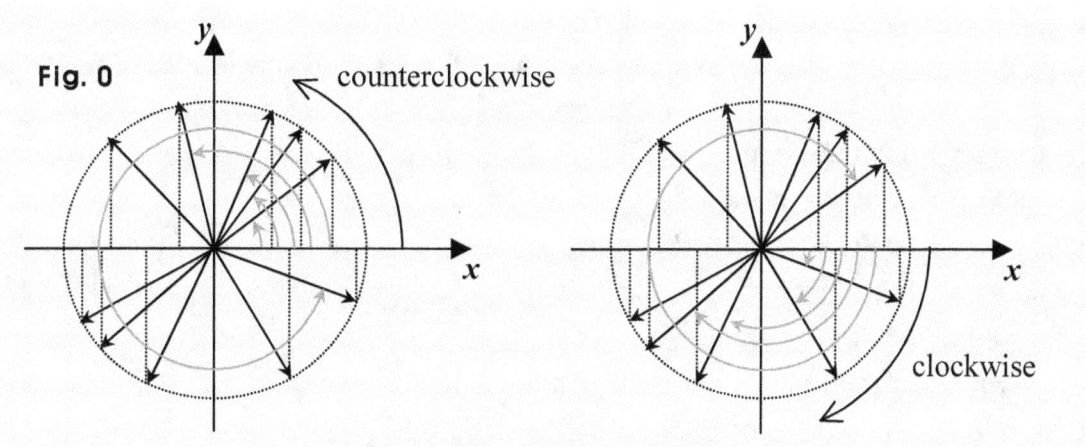

In trigonometry dynamic, the ray turning makes governing angles.
If turning clockwise, it makes angles negative, and if counterclockwise, it makes angles positive. And of course, if no turning, the angle made is 0.
So a governing angle can be $0°$ or any angle positive or negative.

In each right triangle, the ray is of length 1, and is the hypotenuse.
And (x, y) is the terminal point, so x is the adjacent, and y is the opposite.
And thus, assuming θ is a governing angle, we get: $\sin \theta = y$, $\cos \theta = x$, and $\tan \theta = y/x$.

So if the ray is in the first quadrant, all the three trig-ratios > 0, since x and y both > 0.
In the second, only the sines > 0, since $x < 0$, and $y > 0$.
In the third, only the tangents > 0, since $x < 0$, and $y < 0$.
And in the fourth, only the cosines > 0, since $x > 0$, and $y < 0$.

0. Find the value of $\cos \{n\pi + (-1)^n(\pi/6)\}$ where n is an integer.

1. Find the value of $\cos \{n\pi/2 + (-1)^n(\pi/6)\}$ where n is an integer.

Suggestions or Solutions
To the **Problem** in the Example **0**

Find the value of cos $\{n\pi + (-1)^n(\pi/6)\}$ where n is an integer.

The governing angle depends on the value of n.
So let's first, put some values into n beginning with 0, and see how the angle changes.

- $n = 0 \Rightarrow n\pi + (-1)^n(\pi/6) = \pi/6$.
- $n = 1 \Rightarrow n\pi + (-1)^n(\pi/6) = \pi - \pi/6 = 5\pi/6$.
- $n = 2 \Rightarrow n\pi + (-1)^n(\pi/6) = 2\pi + \pi/6$.
- $n = 3 \Rightarrow n\pi + (-1)^n(\pi/6) = 3\pi - \pi/6 = 2\pi + 5\pi/6$.
- $n = 4 \Rightarrow n\pi + (-1)^n(\pi/6) = 4\pi + \pi/6$.
- $n = 5 \Rightarrow n\pi + (-1)^n(\pi/6) = 5\pi - \pi/6 = 4\pi + 5\pi/6$.
- $n = 6 \Rightarrow n\pi + (-1)^n(\pi/6) = 6\pi + \pi/6$.

Now, we can notice that if $n = 2k$ for k integer ≥ 0, the angles are no different from each other.

Fig. 0.0

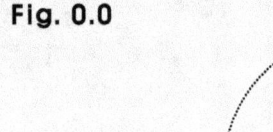

So if $n = 2k$ for k integer ≥ 0, all the rays get the same spot in the first quadrant.

And thus, if $n = 2k$ for k integer ≥ 0, the cosine of $\{n\pi + (-1)^n(\pi/6)\}$ is the same, and is the same as **cos $\pi/6$**, which is $\frac{\sqrt{3}}{2}$.

So if $n = 2k$ for k integer ≥ 0, cos $\{n\pi + (-1)^n(\pi/6)\} = $ cos $\pi/6 = \frac{\sqrt{3}}{2}$.

Next, we can notice that if $n = 2k + 1$ for k integer ≥ 0, the angles are again, no different from each other.

$n = 1 \Rightarrow n\pi + (-1)^n(\pi/6) = \pi - \pi/6 = 5\pi/6$.
$n = 3 \Rightarrow n\pi + (-1)^n(\pi/6) = 3\pi - \pi/6 = 2\pi + 5\pi/6$.
$n = 5 \Rightarrow n\pi + (-1)^n(\pi/6) = 5\pi - \pi/6 = 4\pi + 5\pi/6$.

Fig. 0.1

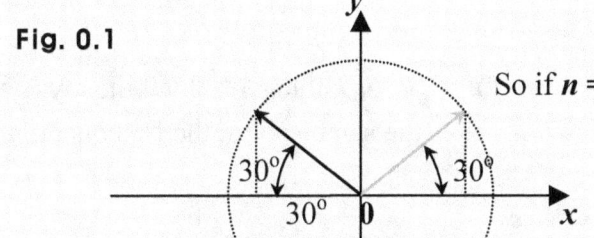

So if $n = 2k + 1$ for k integer ≥ 0, all the rays get the same spot in the second quadrant.

And thus, if $n = 2k + 1$ for k integer ≥ 0, the cosine of $\{n\pi + (-1)^n(\pi/6)\}$ is the same, and is the same as $\cos (5\pi/6)$, which is $-\frac{\sqrt{3}}{2}$.

That's because $\cos (5\pi/6) = -\cos \pi/6$

So if $n = 2k + 1$ for k integer ≥ 0, $\cos \{n\pi + (-1)^n(\pi/6)\} = \cos 5\pi/6 = -\cos \pi/6 = -\frac{\sqrt{3}}{2}$.

So putting together now, the two cases above, we can put them the way below:

If $n = 2k$ for k integer ≥ 0, $\cos \{n\pi + (-1)^n(\pi/6)\} = \frac{\sqrt{3}}{2}$.

And if $n = 2k + 1$ for k integer ≥ 0, $\cos \{n\pi + (-1)^n(\pi/6)\} = -\frac{\sqrt{3}}{2}$.

Next, putting some negative values into n beginning with 0, we get a sequence below:

• $n = 0 \Rightarrow n\pi + (-1)^n(\pi/6) = \pi/6$.

$n = -1 \Rightarrow n\pi + (-1)^n(\pi/6) = -\pi - \pi/6$.

• $n = -2 \Rightarrow n\pi + (-1)^n(\pi/6) = -2\pi + \pi/6$.

$n = -3 \Rightarrow n\pi + (-1)^n(\pi/6) = -3\pi - \pi/6 = -2\pi - \pi - \pi/6$.

• $n = -4 \Rightarrow n\pi + (-1)^n(\pi/6) = -4\pi + \pi/6$.

$n = -5 \Rightarrow n\pi + (-1)^n(\pi/6) = -5\pi - \pi/6 = -4\pi - \pi - \pi/6$.

• $n = -6 \Rightarrow n\pi + (-1)^n(\pi/6) = -6\pi + \pi/6$.

Now, we can notice that if $n = 2k$ for k integer ≤ 0, the angles are no different from each other.

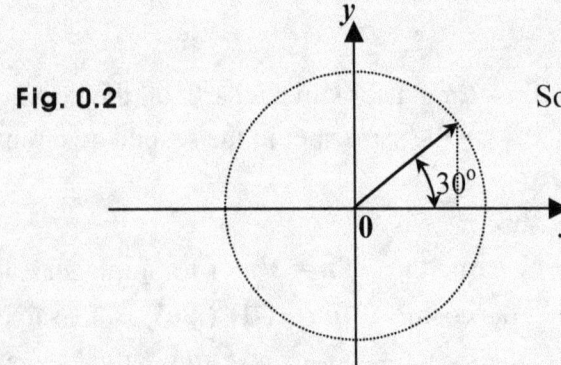

Fig. 0.2

So if $n = 2k$ for k integer ≥ 0, all the rays get the same spot in the first quadrant.

And thus, if $n = 2k$ for k integer ≥ 0, the cosine of $\{n\pi + (-1)^n(\pi/6)\}$ is the same, and is the same as $\cos \pi/6$, which is $\frac{\sqrt{3}}{2}$.

So if $n = 2k$ for k integer ≤ 0, $\cos \{n\pi + (-1)^n(\pi/6)\} = \cos \pi/6 = \frac{\sqrt{3}}{2}$.

Next, we can notice that if $n = 2k - 1$ for k integer ≤ 0, the angles are again, no different from each other.

$n = -1 \Rightarrow n\pi + (-1)^n(\pi/6) = -\pi - \pi/6$.

$n = -3 \Rightarrow n\pi + (-1)^n(\pi/6) = -3\pi - \pi/6 = -2\pi - \pi - \pi/6$.

$n = -5 \Rightarrow n\pi + (-1)^n(\pi/6) = -5\pi - \pi/6 = -4\pi - \pi - \pi/6$.

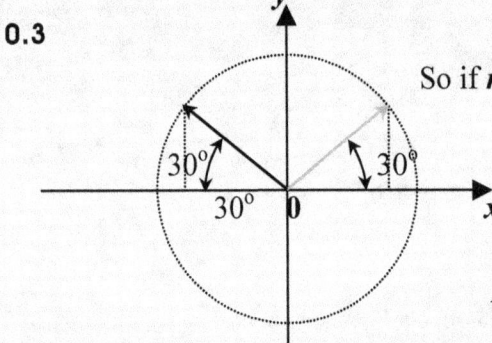

Fig. 0.3

So if $n = 2k - 1$ for k integer ≤ 0, all the rays get the same spot in the second quadrant.

And thus, if $n = 2k - 1$ for k integer ≤ 0, the cosine of $\{n\pi + (-1)^n(\pi/6)\}$ is the same, and is the same as $\cos (-\pi - \pi/6)$, which is $-\frac{\sqrt{3}}{2}$.

That's because $\cos (-\pi - \pi/6) = -\cos \pi/6$

So if $n = 2k - 1$ for k integer ≥ 0, $\cos \{n\pi + (-1)^n(\pi/6)\} = \cos (-\pi - \pi/6) = -\cos \pi/6 = -\frac{\sqrt{3}}{2}$.

So putting together now, the two cases above, we can put them the way below:

If $n = 2k$ for k integer ≤ 0, $\cos \{n\pi + (-1)^n(\pi/6)\} = \frac{\sqrt{3}}{2}$.
And if $n = 2k - 1$ for k integer ≤ 0, $\cos \{n\pi + (-1)^n(\pi/6)\} = -\frac{\sqrt{3}}{2}$.

And also, we have:

If $n = 2k$ for k integer ≥ 0, $\cos \{n\pi + (-1)^n(\pi/6)\} = \frac{\sqrt{3}}{2}$.
And if $n = 2k + 1$ for k integer ≥ 0, $\cos \{n\pi + (-1)^n(\pi/6)\} = -\frac{\sqrt{3}}{2}$.

And taking a closer look at the case where $n = 2k - 1$ for k integer ≤ 0, and the case where $n = 2k + 1$ for k integer ≥ 0, we can just put both together the way below:
$n = 2k + 1$ for k integer. How come?

If $n = 2k + 1$ for k integer ≥ 0, we get: $n = 1, 3, 5, 7, 9$, and so on.

If $n = 2k - 1$ for k integer ≤ 0, we get: $n = -1, -3, -5, -7$, and so forth.

And if $n = 2k + 1$ for k integer, that is, $k = 0, 1, -1, 2, -2, 3, -3, 4, -4$, etc., we get: $n = 1, 3, -1, 5, -3, 7, -5, 9, -7, \ldots$

So putting threads together, we get:

If $n = 2k$ for k integer, $\cos \{n\pi + (-1)^n(\pi/6)\} = \frac{\sqrt{3}}{2}$.
And if $n = 2k + 1$ for k integer, $\cos \{n\pi + (-1)^n(\pi/6)\} = -\frac{\sqrt{3}}{2}$.

And we can put it the way below, too:

If $n = 2k$ for k integer, $\cos \{n\pi + (-1)^n(\pi/6)\} = \frac{\sqrt{3}}{2}$.

And if $n = 2k - 1$ for k integer, $\cos \{n\pi + (-1)^n(\pi/6)\} = -\frac{\sqrt{3}}{2}$.

124

Suggestions or Solutions
To the Problem in the Example 1

Find the value of cos $\{n\pi/2 + (-1)^n(\pi/6)\}$ where n is an integer.

The governing angle depends on the value of **n**.
So let's first, put some values into **n** beginning with 0, and see how the angle changes.

- $n = 0 \Rightarrow n\pi/2 + (-1)^n(\pi/6) = \pi/6.$
  $n = 1 \Rightarrow n\pi/2 + (-1)^n(\pi/6) = \pi/2 - \pi/6 = \pi/3.$
  $n = 2 \Rightarrow n\pi/2 + (-1)^n(\pi/6) = 2\pi/2 + \pi/6 = \pi + \pi/6.$
  $n = 3 \Rightarrow n\pi/2 + (-1)^n(\pi/6) = 3\pi/2 - \pi/6 = \pi + \pi/2 - \pi/6 = \pi + \pi/3.$
- $n = 4 \Rightarrow n\pi/2 + (-1)^n(\pi/6) = 4\pi/2 + \pi/6 = 2\pi + \pi/6.$
  $n = 5 \Rightarrow n\pi/2 + (-1)^n(\pi/6) = 5\pi/2 - \pi/6 = 2\pi + \pi/2 - \pi/6 = 2\pi + \pi/3.$
  $n = 6 \Rightarrow n\pi/2 + (-1)^n(\pi/6) = 6\pi/2 + \pi/6 = 2\pi + \pi + \pi/6.$
  $n = 7 \Rightarrow n\pi/2 + (-1)^n(\pi/6) = 7\pi/2 - \pi/6 = 2\pi + 3\pi/2 - \pi/6 = 2\pi + \pi + \pi/3.$
- $n = 8 \Rightarrow n\pi/2 + (-1)^n(\pi/6) = 8\pi/2 + \pi/6 = 4\pi + \pi/6.$
  $n = 9 \Rightarrow n\pi/2 + (-1)^n(\pi/6) = 9\pi/2 - \pi/6 = 4\pi + \pi/2 - \pi/6 = 4\pi + \pi/3.$
  $n = 10 \Rightarrow n\pi/2 + (-1)^n(\pi/6) = 10\pi/2 + \pi/6 = 4\pi + \pi + \pi/6.$
  $n = 11 \Rightarrow n\pi/2 + (-1)^n(\pi/6) = 11\pi/2 - \pi/6 = 4\pi + 3\pi/2 - \pi/6 = 4\pi + \pi + \pi/3.$
- $n = 12 \Rightarrow n\pi/2 + (-1)^n(\pi/6) = 12\pi/2 + \pi/6 = 6\pi + \pi/6.$

Now, taking a closer look at the sequence of angles above, we can notice that if $n = 4k$
for **k** integer ≥ 0, the angles are no different from each other.

Fig. 1.0

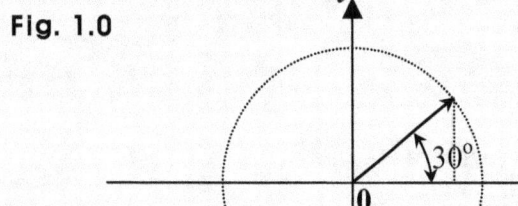

So if $n = 4k$ for **k** integer ≥ 0, all the rays get
the same spot in the first quadrant.

And thus, if $n = 4k$ for **k** integer ≥ 0,
the cosine of $\{n\pi/2 + (-1)^n(\pi/6)\}$ is the same,
and is the same as **cos $\pi/6$**, which is $\frac{\sqrt{3}}{2}$.

So if $n = 4k$ for **k** integer ≥ 0, **cos $\{n\pi/2 + (-1)^n(\pi/6)\} = $ cos $\pi/6 = \frac{\sqrt{3}}{2}$.**

Next, we can notice that if $n = 4k + 1$ for k integer ≥ 0, the angles are again, no different from each other.

$n = 1 \Rightarrow n\pi/2 + (-1)^n(\pi/6) = \pi/2 - \pi/6 = \pi/3$.

$n = 5 \Rightarrow n\pi/2 + (-1)^n(\pi/6) = 5\pi/2 - \pi/6 = 2\pi + \pi/2 - \pi/6 = 2\pi + \pi/3$.

$n = 9 \Rightarrow n\pi/2 + (-1)^n(\pi/6) = 9\pi/2 - \pi/6 = 4\pi + \pi/2 - \pi/6 = 4\pi + \pi/3$.

Fig. 1.1

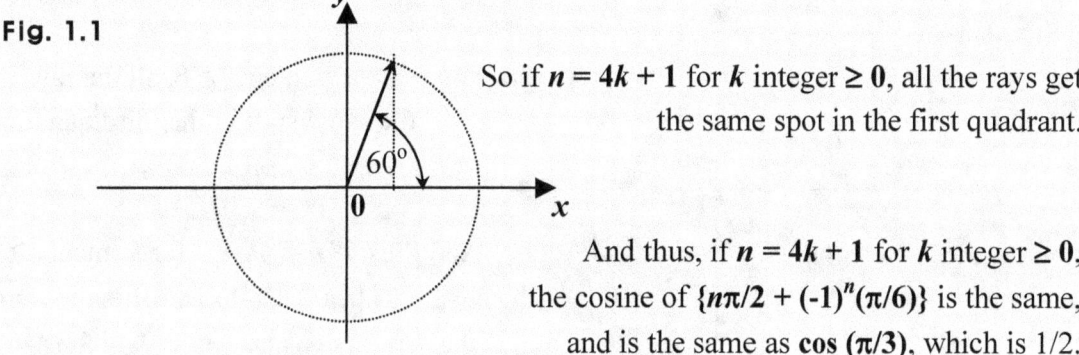

So if $n = 4k + 1$ for k integer ≥ 0, all the rays get the same spot in the first quadrant.

And thus, if $n = 4k + 1$ for k integer ≥ 0, the cosine of $\{n\pi/2 + (-1)^n(\pi/6)\}$ is the same, and is the same as $\cos (\pi/3)$, which is $1/2$.

So if $n = 4k + 1$ for k integer ≥ 0, $\cos \{n\pi/2 + (-1)^n(\pi/6)\} = \cos \pi/3 = 1/2$.

Next, we can notice that if $n = 4k + 2$ for k integer ≥ 0, the angles are again, no different from each other.

$n = 2 \Rightarrow n\pi/2 + (-1)^n(\pi/6) = 2\pi/2 + \pi/6 = \pi + \pi/6$.

$n = 6 \Rightarrow n\pi/2 + (-1)^n(\pi/6) = 6\pi/2 + \pi/6 = 2\pi + \pi + \pi/6$.

$n = 10 \Rightarrow n\pi/2 + (-1)^n(\pi/6) = 10\pi/2 + \pi/6 = 4\pi + \pi + \pi/6$.

Fig. 1.2

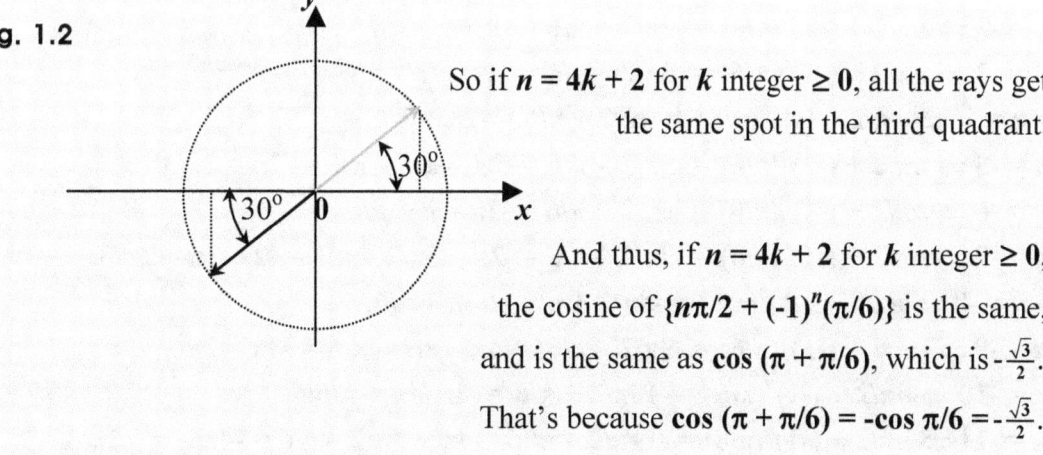

So if $n = 4k + 2$ for k integer ≥ 0, all the rays get the same spot in the third quadrant.

And thus, if $n = 4k + 2$ for k integer ≥ 0, the cosine of $\{n\pi/2 + (-1)^n(\pi/6)\}$ is the same, and is the same as $\cos (\pi + \pi/6)$, which is $-\frac{\sqrt{3}}{2}$.

That's because $\cos (\pi + \pi/6) = -\cos \pi/6 = -\frac{\sqrt{3}}{2}$.

So if $n = 4k + 2$ for k integer ≥ 0, $\cos \{n\pi/2 + (-1)^n(\pi/6)\} = -\cos \pi/6 = -\frac{\sqrt{3}}{2}$.

And next, we can notice that if $n = 4k + 3$ for k integer ≥ 0, the angles are again, no different from each other.

$n = 3 \Rightarrow n\pi/2 + (-1)^n(\pi/6) = 3\pi/2 - \pi/6 = \pi + \pi/2 - \pi/6 = \pi + \pi/3$.

$n = 7 \Rightarrow n\pi/2 + (-1)^n(\pi/6) = 7\pi/2 - \pi/6 = 2\pi + 3\pi/2 - \pi/6 = 2\pi + \pi + \pi/3$.

$n = 11 \Rightarrow n\pi/2 + (-1)^n(\pi/6) = 11\pi/2 - \pi/6 = 4\pi + 3\pi/2 - \pi/6 = 4\pi + \pi + \pi/3$.

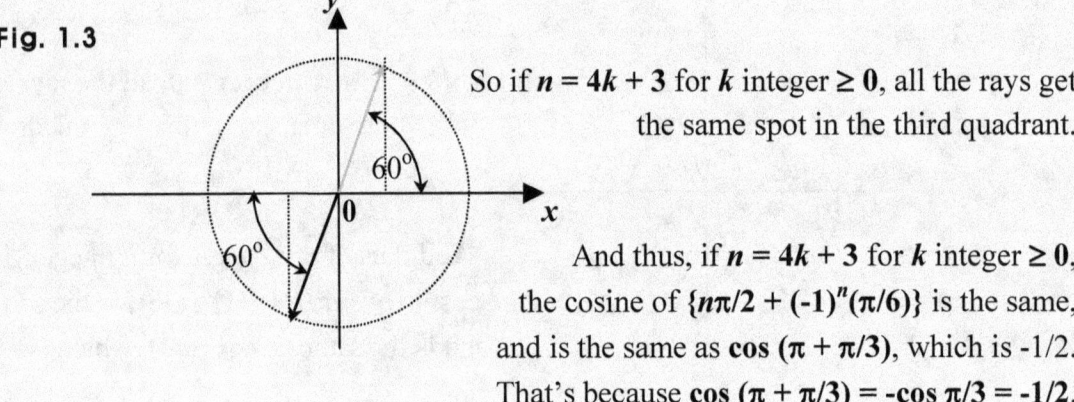

Fig. 1.3

So if $n = 4k + 3$ for k integer ≥ 0, all the rays get the same spot in the third quadrant.

And thus, if $n = 4k + 3$ for k integer ≥ 0, the cosine of $\{n\pi/2 + (-1)^n(\pi/6)\}$ is the same, and is the same as $\cos(\pi + \pi/3)$, which is -1/2. That's because $\cos(\pi + \pi/3) = -\cos \pi/3 = -1/2$.

So if $n = 4k + 3$ for k integer ≥ 0, $\cos\{n\pi/2 + (-1)^n(\pi/6)\} = -\cos \pi/3 = -1/2$.

What if though, k is an integer negative?

Putting some negative values into n beginning with 0, we get a sequence below:

• $n = 0 \Rightarrow n\pi/2 + (-1)^n(\pi/6) = \pi/6$.

$n = -1 \Rightarrow n\pi/2 + (-1)^n(\pi/6) = -\pi/2 - \pi/6 = -2\pi/3$.

$n = -2 \Rightarrow n\pi/2 + (-1)^n(\pi/6) = -2\pi/2 + \pi/6 = -\pi + \pi/6$.

$n = -3 \Rightarrow n\pi/2 + (-1)^n(\pi/6) = -3\pi/2 - \pi/6 = -\pi - \pi/2 - \pi/6 = -\pi - 2\pi/3$.

• $n = -4 \Rightarrow n\pi/2 + (-1)^n(\pi/6) = -4\pi/2 + \pi/6 = -2\pi + \pi/6$.

$n = -5 \Rightarrow n\pi/2 + (-1)^n(\pi/6) = -5\pi/2 - \pi/6 = -2\pi - \pi/2 - \pi/6 = -2\pi - 2\pi/3$.

$n = -6 \Rightarrow n\pi/2 + (-1)^n(\pi/6) = -6\pi/2 + \pi/6 = -2\pi - \pi + \pi/6$.

$n = -7 \Rightarrow n\pi/2 + (-1)^n(\pi/6) = -7\pi/2 - \pi/6 = -2\pi - 3\pi/2 - \pi/6 = -2\pi - \pi - 2\pi/3$.

• $n = -8 \Rightarrow n\pi/2 + (-1)^n(\pi/6) = -8\pi/2 + \pi/6 = -4\pi + \pi/6$.

$n = -9 \Rightarrow n\pi/2 + (-1)^n(\pi/6) = -9\pi/2 - \pi/6 = -4\pi - \pi/2 - \pi/6 = -4\pi - 2\pi/3$.

$n = -10 \Rightarrow n\pi/2 + (-1)^n(\pi/6) = -10\pi/2 + \pi/6 = -4\pi - \pi + \pi/6$.

$n = -11 \Rightarrow n\pi/2 + (-1)^n(\pi/6) = -11\pi/2 - \pi/6 = -4\pi + 3\pi/2 - \pi/6 = -4\pi - \pi - 2\pi/3$.

• $n = -12 \Rightarrow n\pi/2 + (-1)^n(\pi/6) = -12\pi/2 + \pi/6 = -6\pi + \pi/6$.

Now, taking a closer look at the sequence of angles above, we can notice again, if $n = 4k$ for k integer ≤ 0, the angles are no different from each other.

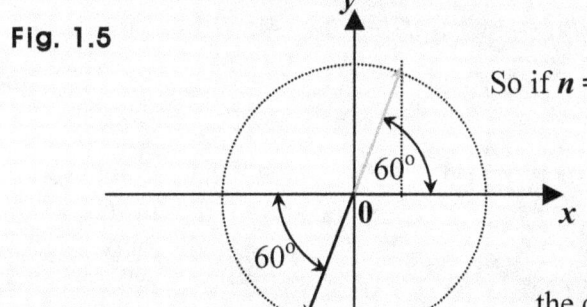

Fig. 1.4

So if $n = 4k$ for k integer ≤ 0, all the rays get the same spot in the first quadrant.

And thus, if $n = 4k$ for k integer ≤ 0, the cosine of $\{n\pi/2 + (-1)^n(\pi/6)\}$ is the same, and is the same as $\cos \pi/6$, which is $\frac{\sqrt{3}}{2}$.

So if $n = 4k$ for k integer ≤ 0, $\cos \{n\pi/2 + (-1)^n(\pi/6)\} = \frac{\sqrt{3}}{2}$.

Next, we can notice that if $n = 4k - 1$ for k integer ≤ 0, the angles are again, no different from each other.

$n = -1 \Rightarrow n\pi/2 + (-1)^n(\pi/6) = -\pi/2 - \pi/6 = -2\pi/3$.

$n = -5 \Rightarrow n\pi/2 + (-1)^n(\pi/6) = -5\pi/2 - \pi/6 = -2\pi - \pi/2 - \pi/6 = -2\pi - 2\pi/3$.

$n = -9 \Rightarrow n\pi/2 + (-1)^n(\pi/6) = -9\pi/2 - \pi/6 = -4\pi - \pi/2 - \pi/6 = -4\pi - 2\pi/3$.

Fig. 1.5

So if $n = 4k - 1$ for k integer ≤ 0, all the rays get the same spot in the third quadrant.

And thus, if $n = 4k - 1$ for k integer ≥ 0, the cosine of $\{n\pi/2 + (-1)^n(\pi/6)\}$ is the same, and is the same as $-\cos (\pi/3)$, which is $-1/2$.

So if $n = 4k - 1$ for k integer ≤ 0, $\cos \{n\pi/2 + (-1)^n(\pi/6)\} = -\cos \pi/3 = -1/2$.

Next, we can notice that if $n = 4k - 2$ for k integer ≤ 0, the angles are again, no different from each other.

$n = -2 \Rightarrow n\pi/2 + (-1)^n(\pi/6) = -2\pi/2 + \pi/6 = -\pi + \pi/6$.

$n = -6 \Rightarrow n\pi/2 + (-1)^n(\pi/6) = -6\pi/2 + \pi/6 = -2\pi - \pi + \pi/6$.

$n = -10 \Rightarrow n\pi/2 + (-1)^n(\pi/6) = -10\pi/2 + \pi/6 = -4\pi - \pi + \pi/6$.

Fig. 1.6

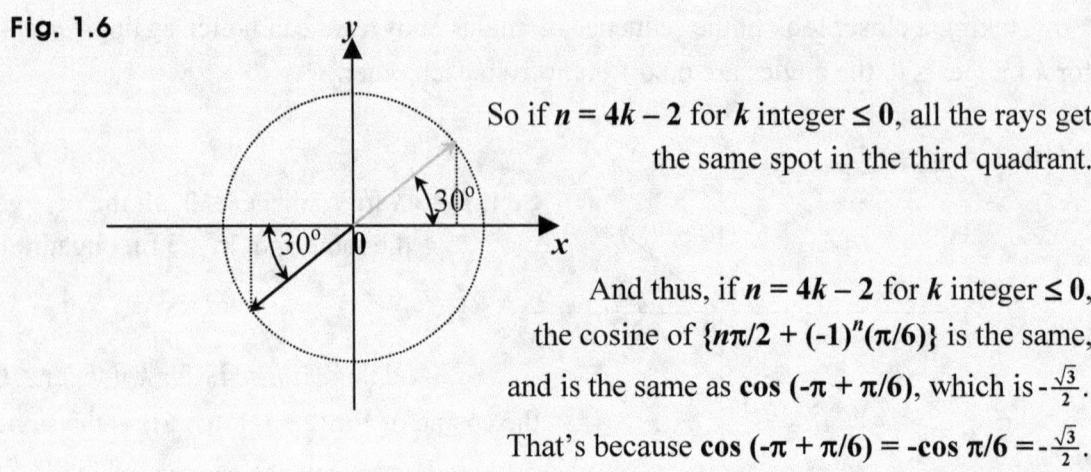

So if $n = 4k - 2$ for k integer ≤ 0, all the rays get the same spot in the third quadrant.

And thus, if $n = 4k - 2$ for k integer ≤ 0, the cosine of $\{n\pi/2 + (-1)^n(\pi/6)\}$ is the same, and is the same as $\cos(-\pi + \pi/6)$, which is $-\frac{\sqrt{3}}{2}$.

That's because $\cos(-\pi + \pi/6) = -\cos \pi/6 = -\frac{\sqrt{3}}{2}$.

So if $n = 4k - 2$ for k integer ≤ 0, $\cos\{n\pi/2 + (-1)^n(\pi/6)\} = -\cos \pi/6 = -\frac{\sqrt{3}}{2}$.

And next, we can notice that if $n = 4k - 3$ for k integer ≤ 0, the angles are again, no different from each other.

$n = -3 \Rightarrow n\pi/2 + (-1)^n(\pi/6) = -3\pi/2 - \pi/6 = -\pi - \pi/2 - \pi/6 = -\pi - 2\pi/3$.

$n = -7 \Rightarrow n\pi/2 + (-1)^n(\pi/6) = -7\pi/2 - \pi/6 = -2\pi - 3\pi/2 - \pi/6 = -2\pi - \pi - 2\pi/3$.

$n = -11 \Rightarrow n\pi/2 + (-1)^n(\pi/6) = -11\pi/2 - \pi/6 = -4\pi + 3\pi/2 - \pi/6 = -4\pi - \pi - 2\pi/3$.

Fig. 1.7

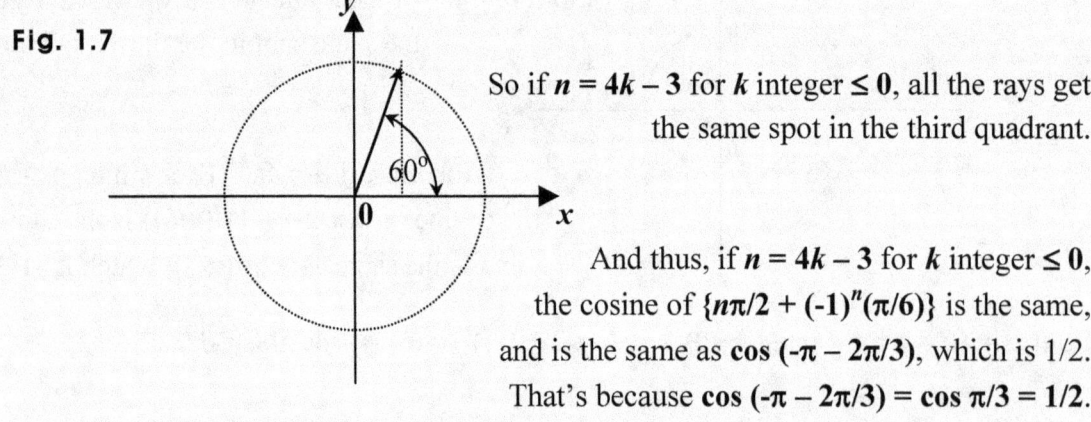

So if $n = 4k - 3$ for k integer ≤ 0, all the rays get the same spot in the third quadrant.

And thus, if $n = 4k - 3$ for k integer ≤ 0, the cosine of $\{n\pi/2 + (-1)^n(\pi/6)\}$ is the same, and is the same as $\cos(-\pi - 2\pi/3)$, which is $1/2$.

That's because $\cos(-\pi - 2\pi/3) = \cos \pi/3 = 1/2$.

Thus, if $n = 4k - 3$ for k integer ≤ 0, $\cos\{n\pi/2 + (-1)^n(\pi/6)\} = \cos \pi/3 = 1/2$.

So putting together now, the case where k is greater than or equal to 0, and the case where k is less than 0, we can put them the way below:

If $n = 4k$ for k integer, $\cos \{n\pi/2 + (-1)^n(\pi/6)\} = \frac{\sqrt{3}}{2}$.

If $n = 4k + 1$ for k integer ≥ 0, $\cos \{n\pi/2 + (-1)^n(\pi/6)\} = 1/2$.

If $n = 4k - 3$ for k integer ≤ 0, $\cos \{n\pi/2 + (-1)^n(\pi/6)\} = 1/2$.

If $n = 4k + 3$ for k integer ≥ 0, $\cos \{n\pi/2 + (-1)^n(\pi/6)\} = -1/2$.

If $n = 4k - 1$ for k integer ≤ 0, $\cos \{n\pi/2 + (-1)^n(\pi/6)\} = -1/2$.

If $n = 4k + 2$ for k integer ≥ 0, $\cos \{n\pi/2 + (-1)^n(\pi/6)\} = -\frac{\sqrt{3}}{2}$.

And if $n = 4k - 2$ for k integer ≤ 0, $\cos \{n\pi/2 + (-1)^n(\pi/6)\} = -\frac{\sqrt{3}}{2}$.

However, as in the case of the sine in the Problem 0 in the Example 2, putting together for instance, the case where $n = 4k + 1$ for k integer ≥ 0, and the case where $n = 4k - 3$ for k integer ≤ 0, we can just put both together the way as follows: $n = 4k + 1$ for k integer.
And the same is true, too, for the other cases where we get the same ratios.
So we can put together all the cases above the way below:

If $n = 4k$ for k integer, $\cos \{n\pi/2 + (-1)^n(\pi/6)\} = \frac{\sqrt{3}}{2}$.

If $n = 4k + 1$ for k integer, $\cos \{n\pi/2 + (-1)^n(\pi/6)\} = 1/2$.

If $n = 4k + 2$ for k integer, $\cos \{n\pi/2 + (-1)^n(\pi/6)\} = -\frac{\sqrt{3}}{2}$.

And if $n = 4k + 3$ for k integer, $\cos \{n\pi/2 + (-1)^n(\pi/6)\} = -1/2$.

And of course, we can put together all the cases above the way below, too:

If $n = 4k$ for k integer, $\cos \{n\pi/2 + (-1)^n(\pi/6)\} = \frac{\sqrt{3}}{2}$.

If $n = 4k - 1$ for k integer ≤ 0, $\cos \{n\pi/2 + (-1)^n(\pi/6)\} = -1/2$.

If $n = 4k - 2$ for k integer ≤ 0, $\cos \{n\pi/2 + (-1)^n(\pi/6)\} = -\frac{\sqrt{3}}{2}$.

If $n = 4k - 3$ for k integer ≤ 0, $\cos \{n\pi/2 + (-1)^n(\pi/6)\} = 1/2$.

So putting the threads together, we want to consider the four cases where $n = 4k$, $4k + 1$, $4k + 2$, and $4k + 3$ for k is an integer.

9. Trigonometric Functions

$$y = f(x) = \sin x \qquad y = g(x) = \cos x \qquad y = h(x) = \tan x$$

So what is a trigonometric function?

In trigonometry dynamic, we can come up with a set of tools, called *trigonometric functions*, often called *trig-functions* for short. And a trig-function is a function of an angle, and thus, takes an angle as an input.   What angle though?

It is a governing angle. So for instance, in a trig-function $y = f(x) = \sin x$, the input variable x takes a governing angle, which is the amount of angle made by the terminal ray turning about the origin in the x-y plane.   So let's now first, get back to the ray turning in the x-y plane, and then, see how it is the case.

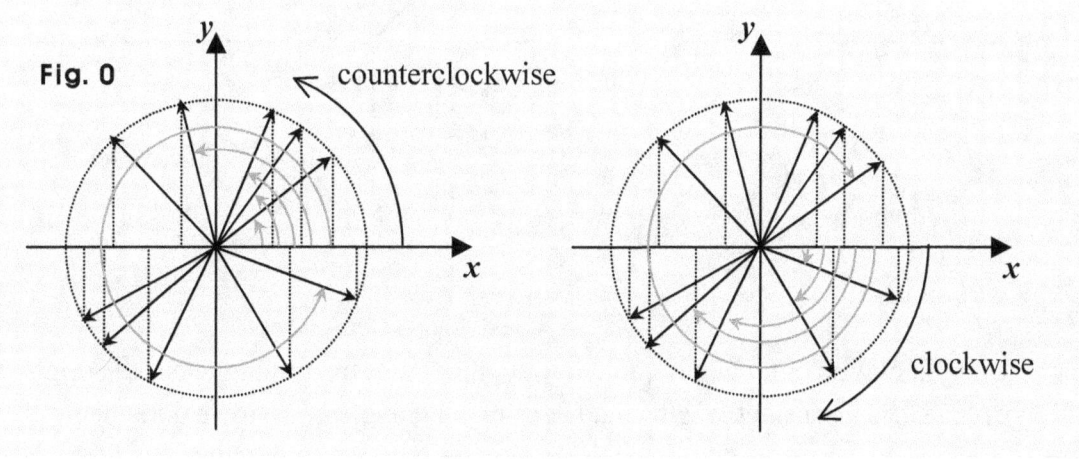

Fig. 0

The ray turning has a length finite, and does not change its length while turning. It keeps changing though, its direction, while turning, of course. And we use an angle to indicate a direction.

So the **ray turning** has a **constant length** and an **angle** that keeps **changing**.

And in the ray turning, the endpoint at the arrowhead can be called the terminal point, and the other endpoint can be called the initial point.
So we can say that the ray in the figure above is turning about the origin in the *x-y* plane, keeping its initial point at the origin, and keeping its length constant.

Assuming now, in the figure above, the terminal point is *(x, y)*, we can see a lot of right triangles. In each, the length of the ray is the length of the hypotenuse, that is, the ray is the hypotenuse, *x* is (the length of) the adjacent, and *y* is (the length of) the opposite.

In short, the **ray** is the **hypotenuse**, *x* is the **adjacent**, and *y* is the **opposite**.

Let's now get back to the ray turning.

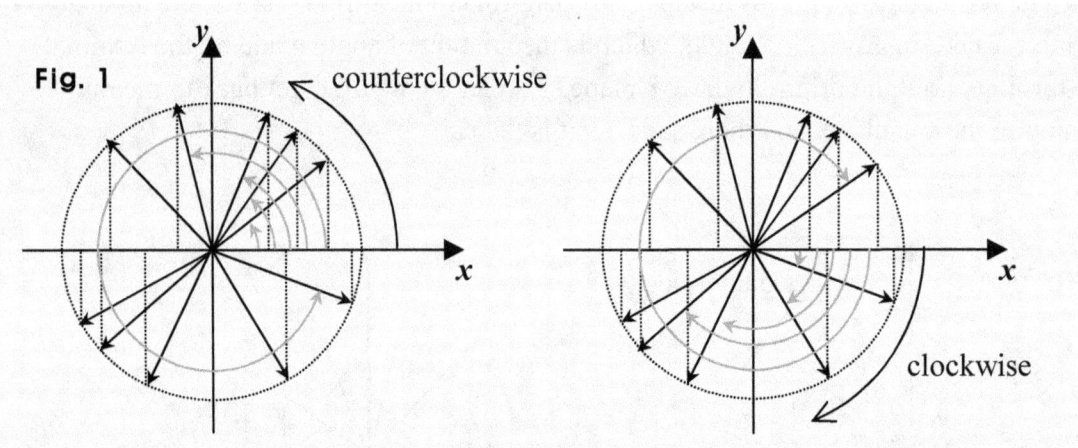

Either counterclockwise or clockwise, in one way, the ray keeps turning about the origin in the *x-y* plane, has a constant length, and keeps making angles.   Why making angles?

Each and every moment, it makes an angle, because an amount of turning is an angle.

And we can have the ray make as many complete turns as necessary either clockwise or counterclockwise. So assuming θ is the angle made, we can put it the way below:

$-360°·n \leq \theta \leq 360°·n \Leftrightarrow -2n\pi \leq \theta \leq 2n\pi$, where n is an integer, and π is 3.141592…, which is a number. That is, we can get: $-\infty < \theta < \infty$, where ∞ indicates infinity.

Thus, θ can be all real numbers, that is, all angles: positive, 0, and negative.

And the ray can be taken as the hypotenuse in each of those right triangles stated above. Then, since the ray is the hypotenuse, the hypotenuse has a constant length and an angle that keeps changing.   What then, is the governing angle?

The angle that the hypotenuse has, that is, the angle that the ray has made is the governing angle in each of the right triangles.

So what's important in the dynamic trigonometry is not only the fact that we can use all angles for trig-ratios but also the fact that the hypotenuse is constant, and another fact that the governing angle is the angle made by the ray turning.

Is the governing angle then, not the angle between the adjacent and the hypotenuse?

Suppose θ is the angle made by the ray.
Then, θ is the governing angle, of course, and only if **$-90° < \theta < 90°$**, θ is the angle between the adjacent and the hypotenuse; otherwise, it's not.

That's not what's really important though. What matters is how governing angles get made, that is, what makes governing angles, since those angles determine trig-ratios.

So what makes governing angles?

134

The ray makes the angles. While turning, each and every moment, the ray makes a governing angle, which is the amount of turning the ray makes.

So what matters is where in the *x-y* plane, the ray turning is at the moment when we need to get the governing angle.

Finding or specifying the position, we can get the amount of angle made by the ray, together with the adjacent and the opposite, of course.

And in trigonometry dynamic, we can say that we work with a right triangle said to be transcendental. A right triangle transcendental is a right triangle made of the ray turning, and two projections.
One is a projection of the ray on the *y*-axis, and the other is the projection on the *x*-axis.

So the two projections are perpendicular to each other. And in fact, we use as the two the coordinates of the terminal point in the ray.

And thus, at any moment while the ray is turning, we can make a right triangle, where the hypotenuse is the ray, the opposite is the projection on the *y*-axis, that is, the *y*-coordinate at the terminal point, and the adjacent is the projection on the *x*-axis, that is, the *x*-coordinate at the terminal point.

And depending on the amount of angle made by the ray, the amounts of the projections get determined. In short, the angle made determines the projections. So what makes such a right triangle is in fact, the ray turning and the amount of angle made by the ray.

In sum, the angle made by the ray determines a right triangle, and thus, determines the opposite and the adjacent. And using the angle made, we can get the trig-ratios.

So each angle made by the ray is a governing angle.

And when we take trig-ratios in a right triangle transcendental, we take the ratios the way we take them in a right triangle normal, too.

So it is always the case where the sine is: the opposite over the hypotenuse, the cosine is the adjacent over the hypotenuse, and the tangent is: the opposite over the adjacent.

However, the governing angle does not have to be the angle between the hypotenuse and the adjacent in the right triangle where the ray is the hypotenuse.

The governing angle is simply the amount of turning the ray has made.

So for instance, if the ray has made one and a sixth of a complete turn counterclockwise, the governing angle is: $(360 + 360/6)^{\circ} = 420^{\circ}$, which is: $2\pi + 2\pi/6 = 7\pi/3$.

Now, in dynamic trigonometry, we often work with functions called trig-functions, of each of which, each input is a governing angle, which is made by the ray turning in the *x-y* plane.

So in the trig-function, too, where $f(x) = \sin x$, the input variable x gets a governing angle, made by the ray at each and every moment while turning as well as at rest before turning.   What then, about the outputs?

Each output is a number. What number though?

It is a trig-ratio. It's because x is a governing angle, so a value of $\sin x$ is a trig-ratio, which is a ratio, which is a number.
So for each value of x, which is a governing angle, $f(x)$ is a number called a trig-ratio.

And in trig-functions, we have three basic ones.
One is a sine function, which is $\sin x$. Another is a cosine function, which is $\cos x$.
And the other is a tangent function, which is $\tan x$.

And usually, we just call <u>governing angles</u>, <u>angles</u>, and use numbers as <u>angles in radian</u>.

So for instance, assuming f is a sine function, and is: $\sin x$, and the domain is a set of all angles, we say that the domain is a set of all real numbers, and can put the trig-function f the way as follows: $y = f(x) = \sin x$ for <u>x real</u>, or simply this way, too: $y = f(x) = \sin x$.

What then, about the range?

The range is a set of all numbers from -1 to 1, and each of the numbers is a trig-ratio, which is an output, which is the value of $f(x)$, which is the value of y.
So the range of the sine function f can be put this way, too: $-1 \leq y \leq 1$, or $|y| \leq 1$.

Note however, if the domain is not a set of all real numbers, the range can be other than the one above.

And the sine function f above can be called the prototype, which is thus, in the most basic form. And assuming F is a sine function, too, and using a general form, we can put the function F the way below:

$$y = F(x) = A \cdot \sin \{w(x + a)\} + b \text{ for } x \text{ real, where } A, w, a, \text{ and } b \text{ are constant.}$$

And we can just put it this way, too: $y = F(x) = A \cdot \sin w(x + a) + b$ for x real.

Or more simply this way, too: $y = F(x) = A \cdot \sin w(x + a) + b$.

Note that though, $w(x + a)$ represents an angle, but A and b represent a number each.

And the function F is defined for x real, so the domain is a set of all angles, that is, a set of all real numbers.

What then, about the range?

The range is a set of all numbers from $-|A| + b$ to $|A| + b$, and is the set of all the outputs.

So the range of F can be put this way, too: $-|A| + b \leq y \leq |A| + b$.

And we call $|A|$ the *amplitude*.

And $|w|$ is called the *frequency*, $\frac{2\pi}{|w|}$ is the *period*, and a is called the *phase*.

• And the same is true for cosine functions, too.

So for instance, assuming g is a cosine function, and is: **cos x**, and the domain is a set of all angles, we say that the domain is a set of all real numbers, and can put the function g the way as follows: $y = g(x) = \cos x$ for x <u>real</u>, or simply this way, too: $y = g(x) = \cos x$.

What then, about the range?

The range is a set of all numbers from -1 to 1, and each of the numbers is a trig-ratio, which is an output, which is the value of $g(x)$, which is the value of y.
So the range of the cosine function g can be put this way, too: **$-1 \le y \le 1$**, or $|y| \le 1$.

As in the case of the sine function f however, if the domain is not a set of all real numbers, it can be the case where the range is not the one above.

And the cosine function g above can be called the prototype, which is thus, in the most basic form. And assuming G is a cosine function, too, and using a general form, we can put the function G the way below:

$$y = G(x) = A \cdot \cos \{w(x + a)\} + b \text{ for } x \text{ real, where } A, w, a, \text{ and } b \text{ are constant.}$$

And of course, we can just put it this way, too: $y = G(x) = A \cdot \cos w(x + a) + b$ for x real.

Or more simply this way, too: $y = G(x) = A \cdot \cos w(x + a) + b$.

And note also that $w(x + a)$ represents an angle, but A and b represent a number each.

And the function G is defined for x real, so the domain is the set of all real numbers.

What then, about the range?

The range is a set of all numbers from $-|A| + b$ to $|A| + b$, and is the set of all outputs.
So the range of G can be put this way, too: $-|A| + b \le y \le |A| + b$.

And we call $|A|$ the *amplitude*, $|w|$ is called the frequency, $\frac{2\pi}{|w|}$ is the period, and a is called the phase, too.

And we will get to the details on A, w, a, and b in one of the sections where we discuss trig-functions.

• What then, about tangent functions?

For instance, assuming h is a tangent function, and is: $\tan x$, we can put the trig-function h the way as follows: $y = h(x) = \tan x$.   What then, about the domain?

In the tangent function h, x takes an angle, too.
Note however, that x cannot take all angles, so the domain of h is not a set of all angles, and thus, is not a set of all real numbers.   Why not though?

The tangent function $\tan x$ cannot be defined if the governing angle x is as follows:

$90° = 1 \cdot 90°$, $270° = 3 \cdot 90°$, $450° = 5 \cdot 90°$, $630° = 7 \cdot 90°$, and so forth.

And the same is true, too, if the angle is: $-90°$, $-270° = -3 \cdot 90°$, $-450° = -5 \cdot 90°$, and such.

And we know: $90° = \pi/2$.

So in short, $\tan x$ cannot be defined for $x = n\pi/2$, where n is an odd integer.

Why not though?

The tangent is: the opposite over the adjacent. So the tangent cannot be defined for an angle that makes the adjacent 0. If for instance, the angle is $90°$, the adjacent is 0.

And the same is true, too, if the angle is: $n\pi/2$, where n is an odd integer.

What then, is the domain of the tangent function $y = h(x) = \tan x$?

The domain of h is a set of all angles less $n\pi/2$ where n is an odd integer, that is, a set of all real numbers other than $n\pi/2$ where n is an integer odd.

So setting: $h(x) = \tan x$, we mean: $h(x) = \tan x$ for $x \neq n\pi/2$ where n is an odd integer.

Note that if the domain is not specified as in the case of $y = h(x) = \tan x$, we just assume that the domain is the set of all possible values that can be the inputs. In other words, the domain is the larges set of values that the input variable can take.

What then, about the range?

Unlike the sine and cosine functions above, the range of the tangent function $\tan x$ is a set of all real numbers, each of which is a trig-ratio, of course.

Note however, if the domain of $\tan x$ is not a set of all real numbers less $n\pi/2$ where n is an odd integer, it can be the case where the range is not a set of all real numbers.

And the tangent function h above can be called the prototype, which is thus, in the most basic form. And assuming H is a tangent function, too, and using a general form, we can put it the way below:

$y = H(x) = A \cdot \tan \{w(x + a)\} + b$, where A, w, a, and b are constant.

And of course, we can just put it this way, too: $y = H(x) = A \cdot \tan w(x + a) + b$.

And note also that $w(x + a)$ represents an angle, but A and b represent a number each.

What then, about the domain?

We know the fact that **tan** θ cannot be defined for $\theta = n\pi/2$ where n is an odd integer.

And also, we know θ is the governing angle in **tan** θ, and in **tan** $w(x + a)$, $w(x + a)$ is the governing angle.

So if $w(x + a) = n\pi/2$ where n is an odd integer, the function H cannot be defined.

In other words, if $w(x + a) \neq n\pi/2$ where n is an odd integer, the function H is defined.

So finding the domain of H, we can get it the way as follows:

$$w(x+a) = \frac{n\pi}{2} \Rightarrow x+a = \frac{n\pi}{2w} \Rightarrow x = \frac{n\pi}{2w} - a. \text{ So if } x \neq \frac{n\pi}{2w} - a, \text{ the function } H \text{ is defined.}$$

And thus, the domain is a set of all angles less $\frac{n\pi}{2w} - a$, where n is an odd integer, that is,

a set of all real numbers other than $\frac{n\pi}{2w} - a$, where n is an odd integer.

What then, about the range?

The range is a set of all real numbers. And as in the case of the sine and cosine functions above, $|w|$ is called the frequency, and a is called the phase, too. However, the period is: not $\frac{2\pi}{|w|}$ but $\frac{\pi}{|w|}$. And we will get to see how it is the case in a separate section.

What about the amplitude?

The tangent function has no amplitude. So A is not the amplitude.
And in fact, we do not apply an amplitude to a tangent function.

And in the next section, we will begin with the sine function prototype, where the amplitude is 1, the phase is 0, and the period is 2π (i.e., $360°$), so the frequency is 1.

And we have another set of three trig-functions, which are the reciprocals of the ones introduced above.

• One is a cosecant function, which is: **csc** *x*, which is the reciprocal of **sin** *x*, so the cosecant function can be just called the reciprocal of the sine function.

• Another is a secant function, which is: **sec** *x*, which is the reciprocal of **cos** *x*, so the secant function can be just called the reciprocal of the cosine function.

• And the other is a cotangent function, which is: **cot** *x*, which is the reciprocal of **tan** *x*, so the cotangent function can be just called the reciprocal of the tangent function.

And we have their inverses, too. The inverse of a function is called an inverse function, and is often just called the inverse, for short. What is an inverse function though?

To begin with, if a function *K* is one-to-one, we can get the inverse.

Then, the domain of the function *K* is the range of the inverse, and the range of the function *K* is the domain of the inverse.

So for instance, if of the function *K*, the domain is *X*, and the range is *Y*, *Y* is the domain of the inverse, and *X* is the range.

And the same is true for the inverse of a trig-function, too.

We know that the domain of a trig-function is a set of angles, and the range is a set of trig-ratios. (Both sets are number sets though, since we just use numbers as angles and ratios are numbers.)

So the domain of the inverse is the set of trig-ratios, and the range is the set of angles. And of course, the trig-function that can have the inverse has to be one-to-one.

And we can define the inverses of the three basic trig-functions the way as follows:

$$y = f(x) = \sin x \Leftrightarrow x = f^{-1}(y) = \sin^{-1} y, \quad y = g(x) = \cos x \Leftrightarrow x = g^{-1}(y) = \cos^{-1} y, \text{ and}$$

$$y = h(x) = \tan x \Leftrightarrow x = h^{-1}(y) = \tan^{-1} y, \text{ where } x \text{ is an angle, and } y \text{ is a trig-ratio.}$$

And by convention, specifying a function, we usually use x as the input variable, and use y as the output variable. And the same is true for inverse functions, too.

And if not ambiguous, we can use a function name over and over.

So to begin with, a sine function $y = f(x) = \sin x$ for $-\pi/2 \leq x \leq \pi/2$ is one-to-one.

So assuming g is the inverse of the sine function f, we get: $y = g(x) = \sin^{-1} x$ for $|x| \leq 1$, where x is a trig-ratio, and y is an angle. And the range of g is: $-\pi/2 \leq y \leq \pi/2$.

Next, a cosine function $y = f(x) = \cos x$ for $0 \leq x \leq \pi$ is one-to-one.

So assuming g is the inverse of the cosine function f, we get: $y = g(x) = \cos^{-1} x$ for $|x| \leq 1$, where x is a trig-ratio, and y is an angle. And the range of g is: $0 \leq y \leq \pi$.

And next, a tangent function $y = f(x) = \tan x$ for $-\pi/2 < x < \pi/2$ is one-to-one.

So assuming g is the inverse of the tangent function f, we get: $y = g(x) = \tan^{-1} x$ for x real, where x is a trig-ratio, and y is an angle. And the range of g is: $-\pi/2 < y < \pi/2$.

If not sure of an inverse function itself, refer to **ALGEBRA EXAMPLES BASIC FUNCTIONS**.

And of course, we will cover the details on each of all the trig-functions above in a separate section.

A. **Sine Functions**

What is a sine function?

It is a function of which each output is a trig-ratio called the sine. So for instance, putting a sine function in the *x-y* coordinate system, we can put it the way below:

$y = f(x) = \sin x$, where *x* is an angle.

So a sine function is a function of an angle, and thus, takes an angle as an input. What angle though?

It is a governing angle. So the input variable *x* takes a governing angle, which is the amount of angle made by the terminal ray turning about the origin in the *x-y* plane.

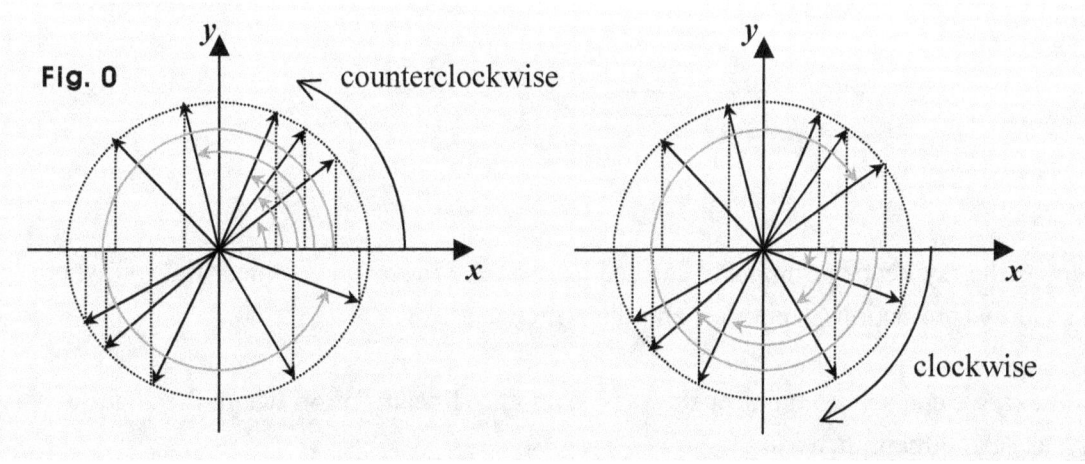

Fig. 0   counterclockwise   clockwise

The ray turning has a length finite, and does not change its length while turning.

And assuming <u>the length of the ray is 1</u>, and the terminal point is *(x, y)*, we can see a lot of right triangles.   What right triangles, though?

In each of the right triangles, the ray is the hypotenuse, *x* is the value of the adjacent, and *y* is the value of the opposite.

 • In short, the ray is the hypotenuse, *x* is the adjacent, and *y* is the opposite.

Suppose now, in the *x-y* plane where the ray is turning counterclockwise, we place two lamps the way below so that the shadow of the ray (the projection) is made on the *x*-axis.

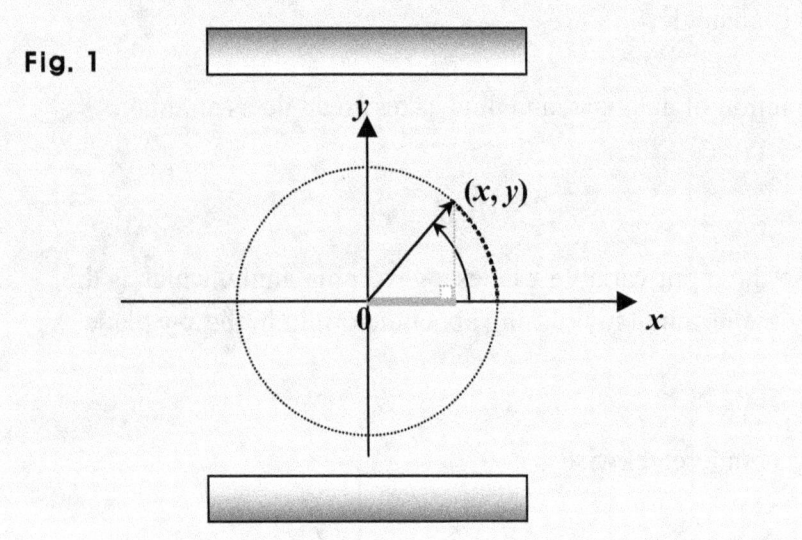

Fig. 1

Then, as the ray keeps turning, we can see from above the *x*-axis, a linear motion where the shadow (projection) decreases, and then, increases.

And next, we can see from below the *x*-axis, another linear motion where the shadow decreases, and then, increases.

And also, we can see that those linear motions keep repeating as the ray keeps turning.

Suppose next, we place two lamps the way below so that the shadow (projection) is made on the *y*-axis.

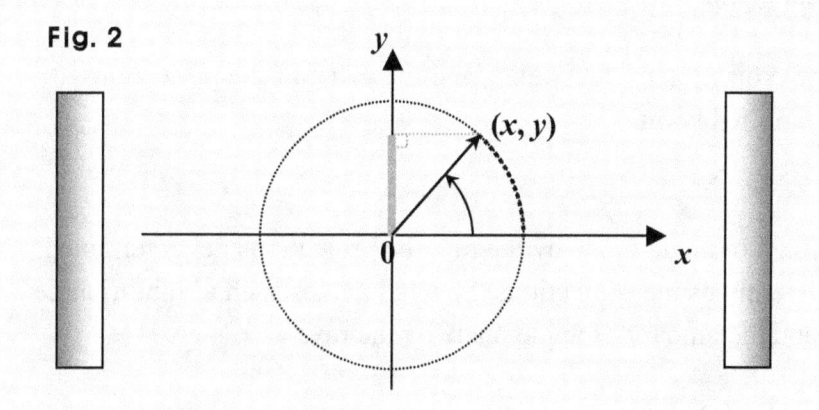

Fig. 2

Then, as the ray keeps turning, we can see from the left of the *y*-axis, a linear motion where the shadow decreases, and then, increases, and also, we can see from the right of the *y*-axis, another linear motion where the shadow decreases, and then, increases.

And also, we can see that such linear motions keep repeating as the ray keeps turning.

So at every moment, assuming the ray is a hypotenuse, the projection on the *x*-axis is the adjacent, and the projection on the *y*-axis is the opposite, we get a right triangle.

And in trigonometry dynamic, we can say that we work with a right triangle said to be transcendental. A right triangle transcendental is a right triangle made of the ray turning, and two projections.

One is a projection of the ray on the *y*-axis, and the other is the projection on the *x*-axis.

So the two projections are perpendicular to each other. And in fact, we use as the two the coordinates of the terminal point in the ray.

And thus, at any moment while the ray is turning, we can make a right triangle.

And in the right triangle:

- The hypotenuse is the ray.

- The opposite is the projection on the y-axis, and the value of the projection is the y-coordinate at the terminal point.

- And the adjacent is the projection on the x-axis, and the value of the projections is the x-coordinate at the terminal point.

And depending on the amount of angle made by the ray, the projections get determined. In short, the angle made determines the projections. So what makes such a right triangle is in fact, the ray turning and the amount of angle made by the ray.

In fact, since the length of the ray is constant, the length of the hypotenuse is constant, too, so the angle made by the ray turning determines a right triangle, because that angle determines the opposite and the adjacent. And we use that angle to get the trig-ratios.

So each angle made by the ray turning is a governing angle

And when taking trig-ratios in a right triangle transcendental, too, we take the ratios the way we take them in a right triangle normal.

So it is always the case where the sine is: the opposite over the hypotenuse, the cosine is the adjacent over the hypotenuse, and the tangent is: the opposite over the adjacent.

- Now, in dynamic trigonometry, we often work with functions called trig-functions. Of each, each input is a governing angle, which is made by the ray turning in the x-y plane.

So in the sine function, $y = f(x) = \sin x$, too, the input variable x gets a governing angle. Usually though, we just call governing angles, angles, and use numbers as angles, since those angles are in radian.

So for instance, assuming f is a sine function, and is: $\sin x$, and the domain is a set of all angles, we say that the domain is a set of all real numbers, and can put the trig-function f the way as follows: $y = f(x) = \sin x$ for x real, or simply this way, too: $y = f(x) = \sin x$.

What then, about outputs?

They are trig-ratios, because in the sine function $y = f(x) = \sin x$, the x gets an angle, so each value of **sin x**, that is, each value of $f(x)$ is a trig-ratio, which is a number.
What then, about the range?

The range is a set of all numbers from -1 to 1, and each of the numbers is a trig-ratio, which is an output, which is the value of $f(x)$, which is the value of y.
So the range of the sine function f can be put this way, too: **-1 ≤ y ≤ 1**, or **|y| ≤ 1**.
How come though?

Let's now, get back to the **x-y** plane where the projection of the ray is made on the **y**-axis.

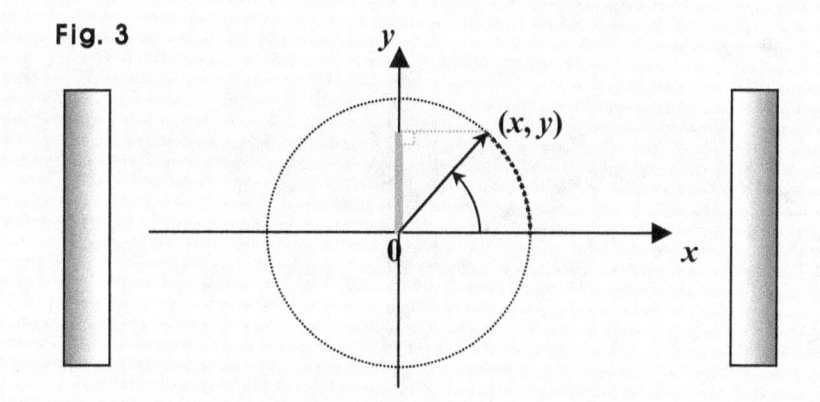

Fig. 3

Suppose that <u>the ray is of length 1</u>, and is placed on the **x**-axis to the right of the origin.

Suppose now that the ray starts turning counterclockwise.

Then, from the right of the **y**-axis, we can see that (the length of) the projection on the **y**-axis starts growing upward from 0.

Even before the ray turns, assuming the ray is a hypotenuse, we can get a right triangle.
What right triangle?

It is a right triangle transcendental, where the hypotenuse is the ray, the adjacent is the projection on the *x*-axis, and the opposite is the projection on the *y*-axis.

Before the ray turns, the ray is at rest on the *x*-axis to the right of the origin, so the projection on the *x*-axis is the ray itself, and the projection on the *y*-axis is 0.

So in this case, (the length of) the adjacent is the same as (the length of) the ray, that is, the hypotenuse, which is 1, and the opposite is 0, which is impossible though, in a right triangle normal. And thus, we can call it a right triangle transcendental.

And we know that the angle made is 0 at that moment, and that by definition, the sine is: the opposite over the hypotenuse.   So we get: **sin 0 = 0/1 = 0**.

Suppose now that the angle made is **θ**, and that the terminal point in the ray is **(x, y)**. What right triangle then, can we get?

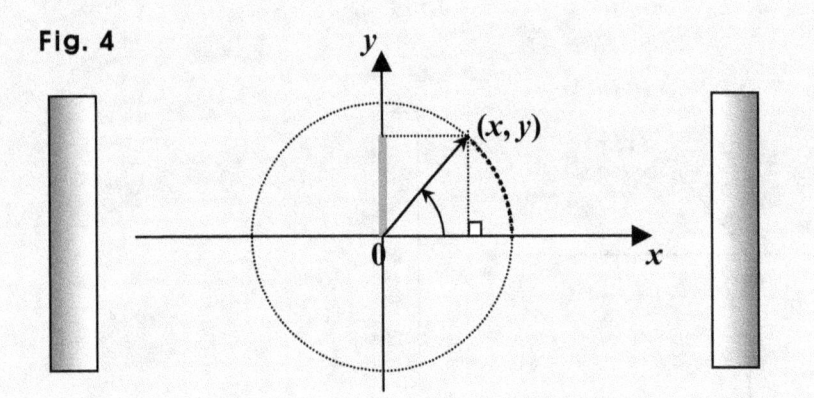

Fig. 4

It is a right triangle transcendental, where the hypotenuse is the ray, the adjacent is *x*, and the opposite is *y*.   What then, is the sine of the angle **θ**, that is, the value of **sin θ**?

We know by definition, the sine is: the opposite over the hypotenuse. And we know the hypotenuse is 1, since the ray is 1, and the opposite is *y*.   So we get: **sin θ = y/1 = y**.

•• So for any angle **θ**, assuming the ray is 1, that is, the hypotenuse is 1, and the terminal point is **(x, y)**, we get: **sin θ = y**, which is of course, the *y*-coordinate at the terminal point.

Let's see now if it is the case checking with some specific angles.

> • Suppose first, that the angle θ is $\pi/2$.   What right triangle then, can we get?

It is a right triangle transcendental, where the hypotenuse is the ray, and the opposite is the projection on the y-axis, which is now the ray itself.   What then, is **sin $\pi/2$?**

It is 1, because both the opposite and the hypotenuse are the same.

We know if θ is $\pi/2$, the ray is on the y-axis, above the origin, of course, the ray is of length 1, and the terminal point is (x, y).   So what is the y-value at (x, y) if θ is $\pi/2$?

It is 1. So the y-value at (x, y) is the value of the opposite, which is 1 when θ is $\pi/2$.

> • Suppose next that the angle θ is π.   What right triangle then, can we get?

It is a right triangle transcendental, where the hypotenuse is the ray, and the opposite is the projection on the y-axis, which is now 0.   What then, is **sin π?**

It is 0, because the opposite is 0.

We know if θ is π, the ray is on the x-axis, to the left of the origin, of course, the ray is of length 1, and the terminal point is (x, y).   So what is the y-value at (x, y) if θ is π?

It is 0. So the y-value at (x, y) is the value of the opposite, which is 0 when θ is π.

> • Suppose now that the angle θ is $3\pi/2$.   What right triangle then, can we get?

It is a right triangle transcendental, where the hypotenuse is the ray, and the opposite is the projection on the *y*-axis, which is now -1.   What then, is **sin 3π/2**?

It is –1, since the opposite is –1, and the hypotenuse is 1.
We know if *θ* is 3π/2, the ray is on the *y*-axis, below the origin, of course, the ray is of length 1, and the terminal point is (*x, y*).   So what is the *y*-value at (*x, y*) if *θ* is 3π/2?

It is –1.   So the *y*-value at (*x, y*) is the value of the opposite, which is -1 when *θ* is 3π/2.

Meanwhile, the ray keeps turning now.

Fig. 5

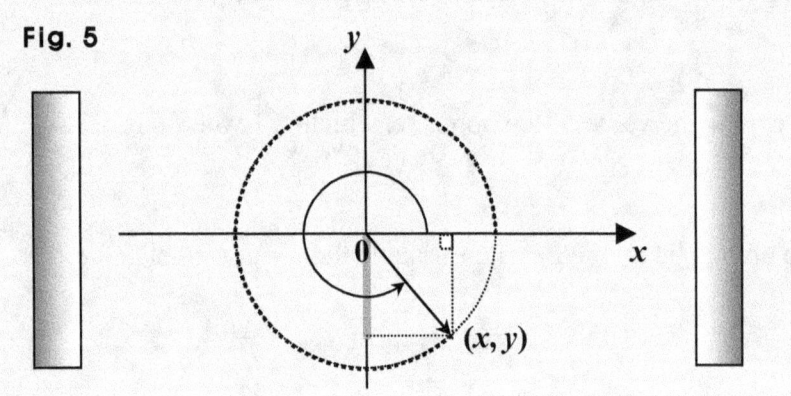

What then, are the maximum and the minimum of the *y*-value at (*x, y*)?

The maximum is 1, and the minimum is –1.
And of course, the maximum and the minimum have to be so, since the terminal point makes a circle of radius 1 centered at the origin if the ray makes a complete turn.

 • Now, as the governing angle *θ* changes from 0 to 2π, what value does **sin *θ*** get?

If the ray is 1, the hypotenuse is 1, so the value of **sin *θ*** is the *y*-value at the terminal point (*x, y*), and *y* covers all the numbers from –1 to 1.
So if **0 ≤ *θ* ≤ 2π**, we get: **-1 ≤ sin *θ* ≤ 1**, that is, |**sin *θ*| ≤ 1**.   What if **2π ≤ *θ* ≤ 4π**?

If $2\pi \leq \theta \leq 4\pi$, we get: $-1 \leq \sin \theta \leq 1$, too, that is, $|\sin \theta| \leq 1$.

And thus, if $0 \leq \theta \leq 2n\pi$, where n is an integer ≥ 1, we get: $|\sin \theta| \leq 1$.

That is, no matter how many complete turns the ray may make, we get: $|\sin \theta| \leq 1$.

Suppose now again, the ray of length 1 is at rest on the *x*-axis to the right of the origin.

- Suppose this time, the ray turns clockwise.

Then, from the right of the *y*-axis, we can see at the beginning, the projection on the *y*-axis starts growing downward from 0.

Suppose now that the angle made is θ, and that the terminal point in the ray is (x, y). What right triangle then, can we get?

It is a right triangle transcendental, where the hypotenuse is the ray, the adjacent is *x*, and the opposite is *y*. What then, is the sine of the angle θ?

We know <u>the sine is: the opposite over the hypotenuse</u>, the hypotenuse is 1, and the opposite is *y*. So we get: $\sin \theta = y/1 = y$. Note that the angle θ is negative now.

- Suppose now that the angle θ is $-\pi/2$. What right triangle then, can we get?

It is a right triangle transcendental, where the hypotenuse is the ray, and the opposite is the projection on the *y*-axis, which is now the ray itself. What then, is $\sin (-\pi/2)$?

It is -1, because the opposite is -1, and the hypotenuse is 1.
We know if θ is $-\pi/2$, the ray is on the *y*-axis, below the origin, of course, the ray is of length 1, and the terminal point is (x, y). So what is the *y*-value at (x, y) if θ is $-\pi/2$?

It is -1. So the *y*-value at (x, y) is the value of the opposite, which is -1 when θ is $-\pi/2$.

152

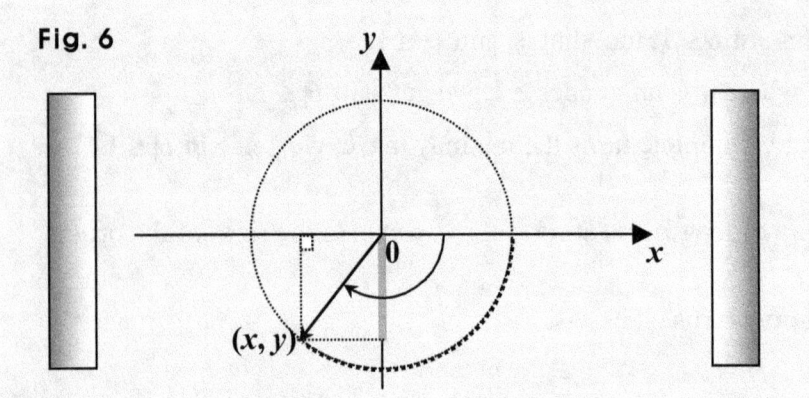

Fig. 6

• Suppose now that the angle θ is -π.   What right triangle then, can we get?

It is a right triangle transcendental, where the hypotenuse is the ray, and the opposite is the projection on the *y*-axis, which is now 0.   What then, is **sin (-π)**?

It is 0 since the opposite is 0.

We know if θ is -π, the ray is on the *x*-axis, to the left of the origin, of course, the ray is of length 1, and the terminal point is (*x, y*).   So what is the *y*-value at (*x, y*) if θ is -π?

It is 0. So the *y*-value at (*x, y*) is the value of the opposite, which is 0 when θ is -π.

• Suppose now that the angle θ is -3π/2.   What right triangle then, can we get?

It is a right triangle transcendental, where the hypotenuse is the ray, and the opposite is the projection on the *y*-axis, which is now 1.   What then, is **sin (-3π/2)**?

It is 1, since the opposite is 1, and the hypotenuse is 1.

We know if θ is -3π/2, the ray is on the *y*-axis, above the origin, of course, the ray is of length 1, and the terminal point is *(x, y)*.   So what is the *y*-value at *(x, y)* if θ is -3π/2?

It is 1.   So the *y*-value at *(x, y)* is the value of the opposite, which is 1 when θ is -3π/2.

And the maximum of the *y*-value at *(x, y)* is 1, and the minimum is –1, since the terminal point makes a circle of radius 1 centered at the origin if the ray makes a complete turn.

• So now, as the governing angle θ changes from 0 to -2π, what value does **sin θ** get?

If the ray is 1, that is, the hypotenuse is 1, the value of **sin θ** is the *y*-value at the terminal point *(x, y)*, and *y* covers all the numbers from –1 to 1.
So if **-2π ≤ θ ≤ 0**, we get: **-1 ≤ sin θ ≤ 1**, that is, **|sin θ| ≤ 1**.   What if **-4π ≤ θ ≤ -2π**?

If **-4π ≤ θ ≤ -2π**, we get: **-1 ≤ sin θ ≤ 1**, too, that is, **|sin θ| ≤ 1**.
And thus, if **2nπ ≤ θ ≤ 0**, where *n* is an integer ≤ **-1**, we get: **|sin θ| ≤ 1**.

So now, putting threads together, if the ray is 1, and *(x, y)* is the terminal point, we get: **sin θ = y** for any angle θ, and get: **|sin θ| ≤ 1**, since |*y*| ≤ **1**.

It's because <u>the sine is: the opposite over the hypotenuse</u>, the hypotenuse is 1, since the ray is 1, and the opposite is the *y*-value at the terminal point *(x, y)*, which makes a circle of radius 1 centered at the origin if the ray makes a complete turn.

••• So using the facts above, we can form a function of the angle θ.   How?

As the angle θ changes, the sine trig-ratio **sin θ** changes.
So we can take as an input, each angle that θ gets, and take as an output, each trig-ratio that **sin θ** produces.

Assuming thus, the function is *f*, the domain is a set of all angles, and *s* is the output variable, we can set: $s = f(\theta) = \sin \theta$.   And we call the function *f* a sine function.

• What then, about the curve of the sine function $s = f(\theta) = \sin \theta$?

Assuming the ray is 1, and the terminal point is (*x, y*), we get: $\sin \theta = y$. So the *y*-value at the terminal point (*x, y*) in the ray that makes the angle θ is the value of $\sin \theta$. Thus, we can get the curve of $\sin \theta$ the way as follows:

Fig. 7

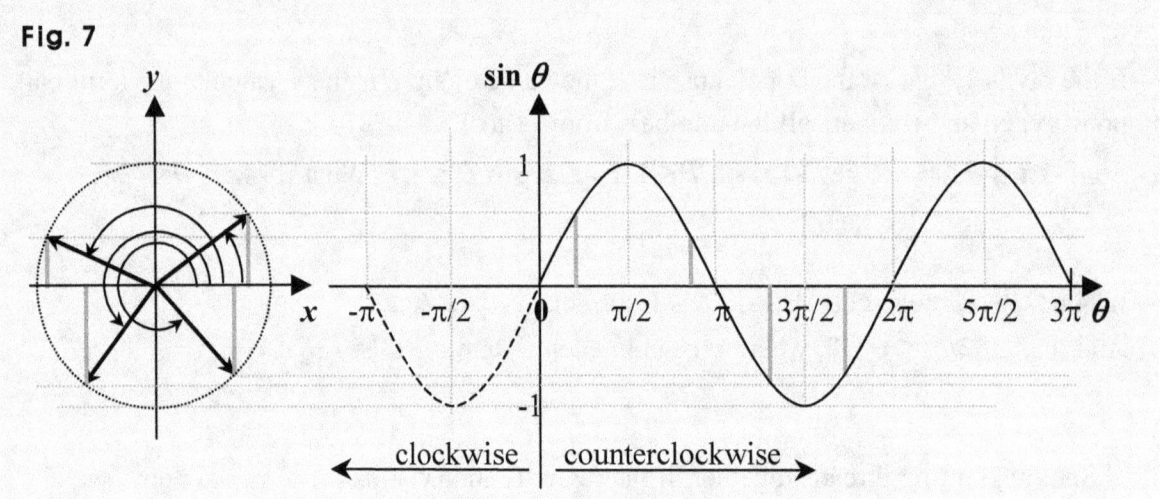

And of course, we can put in the *x-y* system, the function $s = f(\theta) = \sin \theta$.   How?

Replacing *s* with *y*, and θ with *x*, we get: $y = f(x) = \sin x$, which is now in the *x-y* system.

And we can use numbers as angles. And thus, if the domain is a set of all angles, we say that the domain is a set of all real numbers, and we can just set: $y = f(x) = \sin x$ for *x* real.

What if we just set: $y = f(x) = \sin x$?

Then, the domain is assumed to be a set all real numbers. And we can put in the *x-y* plane, the curve of the sine function $y = f(x) = \sin x$ the way below:

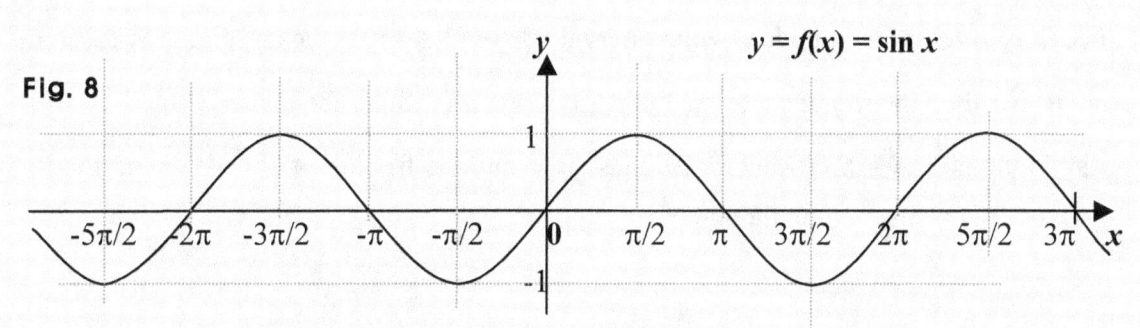

Fig. 8

$$y = f(x) = \sin x$$

The curve above is called the sine curve, and is no other than the curve in **Fig. 7** above.

And we know that the output can be every value from -1 to 1.
So the range is: **-1 ≤ y ≤ 1**, that is, |y| ≤ **1**.

And of course, if the domain is not a set of all real numbers, the range can be other than the one above. Suppose for instance, $y = g(x) = \sin x$ for $0 \le x \le \pi$.
Then, as we can see in the graph above, the range is: **0 ≤ y ≤ 1**.

And the sine function $y = f(x) = \sin x$ can be called the prototype, and is thus, in the most basic form. Assuming F is a sine function, too, where the domain is a set of all real numbers (that is, all angles), and using a general form, we can put it the way below:

$$y = F(x) = A \cdot \sin \{w(x + a)\} + b \text{ for } x \text{ real, where } A, w, a, \text{ and } b \text{ are constant.}$$

And we can just put it this way, too: $y = F(x) = A \cdot \sin w(x + a) + b$ for x real.

Or more simply, this way, too: $y = F(x) = A \cdot \sin w(x + a) + b$.

(Note however, $w(x + a)$ represents an angle, but A and b represent just a number each.)

Then, the range is a set of all numbers from $-|A| + b$ to $|A| + b$.

So the range of F can be put this way, too: $-|A| + b \le y \le |A| + b$.
That is, the curve of F is bounded by the interval where $-|A| + b \le y \le |A| + b$.

Then, $2|A|$ is the width of the curve (or wave), and $|A|$ is called the amplitude.
So the amplitude indicates half the width of the curve (or wave).

What then, is the amplitude of the sine function $y = f(x) = \sin x$?

Since the curve of *f* is bounded by the interval where $-1 \leq y \leq 1$, the amplitude is 1.

And $|w|$ is called the frequency, $\frac{2\pi}{|w|}$ is the period, and *a* is called the phase.

So in the function $y = f(x) = \sin x$, the phase is 0, and the frequency is 1, so the period is 2π (i.e., $360°$). How come though?

The prototype in sine functions can be put this way: $y = f(x) = \sin x$.

And using a general form, we can put a sine function called *F* the way above.

Then, *a* is called the phase, $|w|$ is called the frequency, and $\frac{2\pi}{|w|}$ is the period.

And we can put the prototype into the general form the way below:

$y = f(x) = 1 \cdot \sin 1(x + 0) + 0.$

So the phase is 0, and the frequency is 1, so the period is 2π.

What do we mean by the period though?

Putting in the *x-y* plane, the curve of the sine function *f* above, we can get:

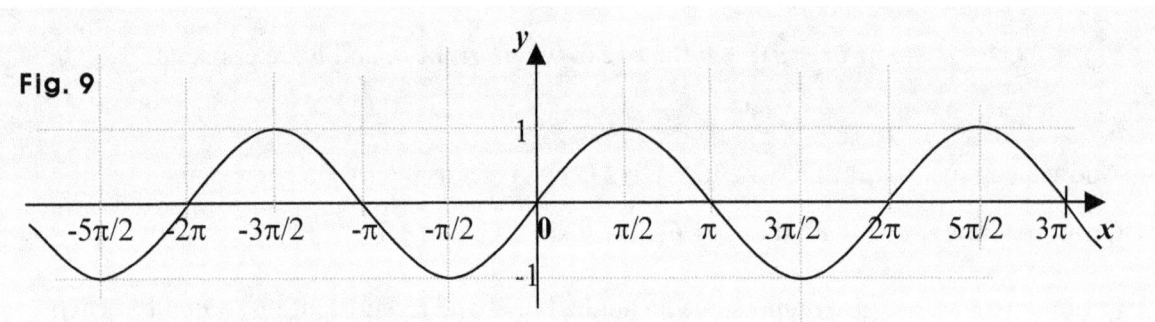

Fig. 9

Then, we can see that a part of the cuve repleats itself.

And the part is in fact, the smallest part repeating itself. What then, is the part?

It is the part from 0 to 2π. So the part repeats itself every 2π interval. And we call such an interval a period. So the interval 2π is the period in the sine function $y = f(x) = \sin x$.

• And thus, we call a sine function a *periodic* function. And of course, the same is true, too, for cosine functions, tangent functions, and other trig-functions as secant functions.

The sine function *f* above is the prototype, and thus, is in the most basic form.

  • So the period **2π** can be called the *basic period* in *sine functions*. In other words:
  • The interval **2π** can be called the *basic interval* in *sine functions*.

And for instance, in the general form, setting *w* to 2, *a* and *b* to 0 each, and *A* to 1, we get a new function where: $y = g(x) = \sin 2x$ for *x* real.

Then, putting in the *x-y* plane, the curve of the sine function *g* above, we get:

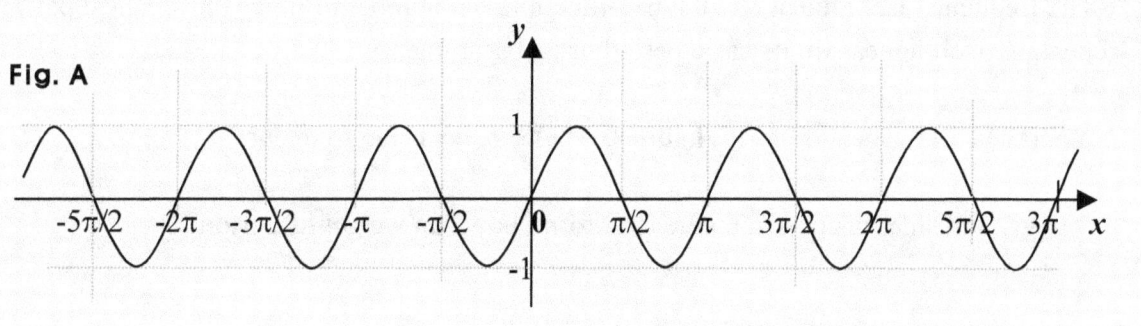

Fig. A

Then again, we can see that a part of the cuve repleats itself.
And the part is of course, the smallest part repeating itself.   What then, is the part?

It is the part from 0 to π.   So the part repeats itself every π interval.

And we call such an interval a period.   So the interval π is the period in the function *g*.

How many times does the smallest part stated above appear in the basic interval 2π?

It appears twice in the basic interval 2π, which is the basic period. Then, we say the frequency is 2 in the function *g*.   What then, do we mean by the frequency?

It is the number of times the smallest part repeating itself appears in the basic inverval 2π, which is the basic period. In short, the frequency is the number of times the smallest part appears in 2π. More precisely though, we want to put the frequency the way below:

- In case of sine functions, the frequency is the number of times the samllest part repeating itself appears in the basic period (or the basic interval), which is 2π.

And the same is true for cosine functions, too. So the basic period (interval) in cosine functions is 2π, too. The basic period however, in tangent functions is not 2π but π.

So in general, we can put the frequency the way below:

- The frequency is, of the curve of a trig-function, the number of times the smallest part repeating itself appears in its basic period or its basic interval.

And thus, in the general form $A{\cdot}\sin w(x + a) + b$, $|w|$ is the frequency.

Then, given a sine function in the general form, how can we get its period?

We know that $|w|$ is the frequency, that is, the number of times the smallest part repeating itself appears in the basic interval (basic period) 2π, and that the interval that fits the smallest part is the period.

So using the frequency $|w|$ and the basic interval 2π, how can we express the period?

We can put the period this way: $\frac{2\pi}{|w|}$. Why not just w but $|w|$ though?

We can have a sine function where $y = h(x) = \sin(-x)$ for x real. Then, the frequency in the sine function h is $|-1|$, simply because a frequency is positive.

And we can put h this way, too: $y = h(x) = -\sin x$ for x real.

It's because: **sin (-x) = -sin x**. Assuming the ray (hypotenuse) is of length 1, we get:

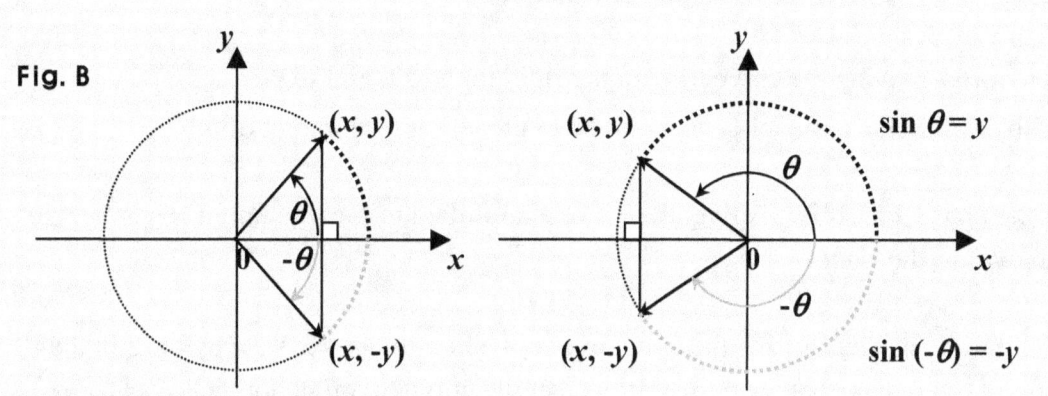

Fig. B

So we get: **sin (-θ) = -sin θ**. And putting in the *x-y* plane, the curve of *h*, we get:

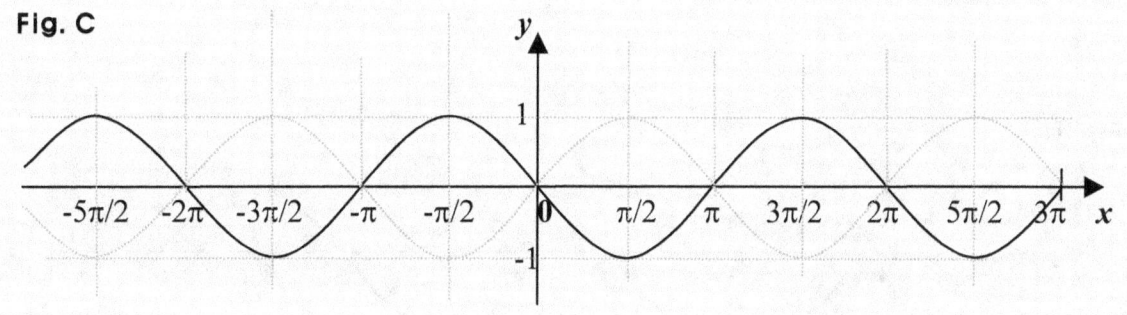

Fig. C

And we can see that the period of *h* is 2π, too.

• And next, what do we mean by the phase?

It can be called a horizontal shift, too.   How come?

Suppose first: *y* = *F*(*x*) = *A*·sin *w*(*x* + *a*) + *b* for *x* real, where *A*, *w*, *a*, and *b* are constant.

Then, *a* is called the phase, |*w*| is called the frequency, and $\frac{2\pi}{|w|}$ is the period.

And next, assuming: *y* = *Q*(*x*) = *A*·sin *wx* + *b* for *x* real, and shifting the curve of the function *Q*(*x*) by −*a* in the direction of the *x*-axis, we get the curve of the function *F* specified above.

So the two curves themselves of **F** and **Q** are the same, and moving the curve of **Q** in the amount of –**a** along the **x**-axis, we get the curve of **F**.

In other words, moving the curve of **Q** to the right in the amount of –**a**.
That is to say that the curve of **Q** gets moved to the left in the amount of **a**.

• So if the <u>phase</u> **a** itself is <u>positive</u>, the curve gets shifted to the <u>left</u>, and if *negative*, the curve moves to the *right*.

Assuming for instance, shifting the curve of $p(x) = \sin x$ for $-\pi \le x \le 3\pi$ by $-\pi/2$ in the direction of the **x**-axis, that is, to the left, we get the curve of a function below:

$q(x) = \sin(x + \pi/2)$ for $-3\pi/2 \le x \le 5\pi/2$.

So putting in a graph the two curves, that is, the curves of **p** and **q**, we get:

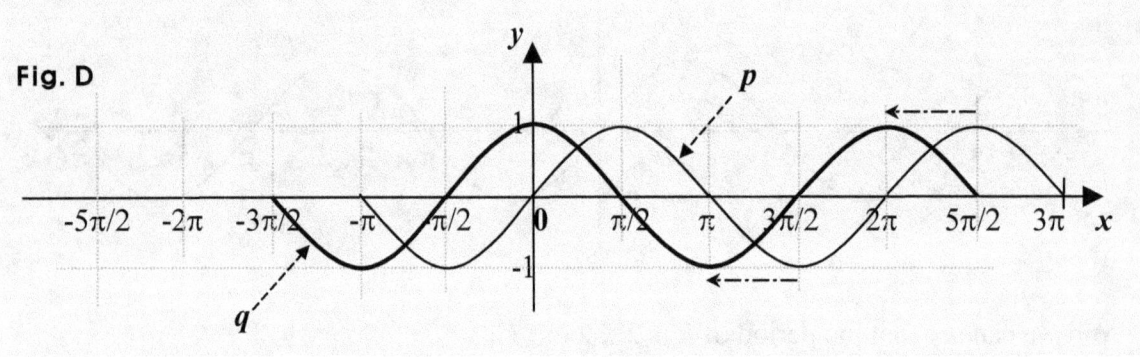

Fig. D

And thus, the curve of **p** gets moved to the left in the amount of $\pi/2$. And the new curve is the curve of **q**. So we can call the phase a *horizontal* shift, too.

What then, about the constant **b** in $y = Q(x) = A \cdot \sin wx + b$?

It can be called a *vertical* shift.

So the constant **b** makes a curve move vertically and in the amount of **b**.

In other words, it makes a curve move in the amount of **b** in the direction of the **y**-axis.

So for instance, assuming: $y = V(x) = A \cdot \sin w(x + a)$ for x real, and shifting the curve of the function $V(x)$ by b in the direction of the y-axis, we can get the curve of the function $y = F(x) = A \cdot \sin w(x + a) + b$ for x real.

So the two curves themselves of F and V are the same, and moving the curve of V in the amount of b along the y-axis, we get the curve of F.

And if b is positive, the curve is shifted upward, and if negative, it moves downward.

Assuming for instance, shifting by 1 in the direction of the y-axis, that is, upward, the curve of $u(x) = \sin x$ for x real, we get the curve of a function as follows:

$v(x) = \sin x + 1$ for x real.

Putting therefore, the two curves in a graph, we get:

Fig. E

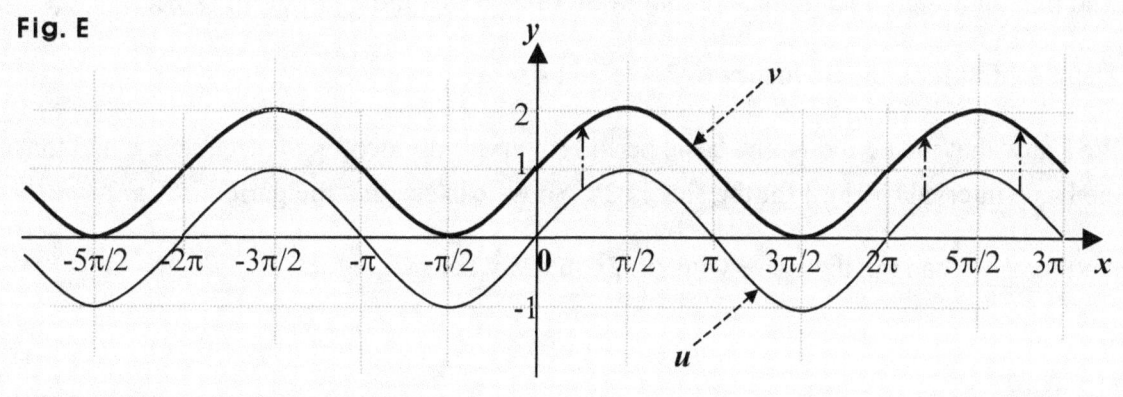

So the curve of u gets moved upward in the amount of 1.
And the new curve is the curve of v. So we can call the b the amount of a vertical shift. In short, b is a vertical shift.

And formally, we call such shifting a *transformation*, too.
More specifically, it is called a *parallel transformation*, because it forces a curve to *translate*.

If not sure of such a transformation, refer to **ALGEBRA EXAMPLES GRAPH OPERATIONS**.

162

And let's now put in a graph, for instance, the curve of a sine function below:

$y = S(x) = $ -2 sin (-2x + π) + 1 for x real.

To begin with, we have: **sin (-θ) = -sin θ.**

So we can get: **sin (-2x + π) = sin {-(2x − π)} = -sin (2x − π).**

Thus, we get: **-2 sin (-2x + π) + 1 = 2 sin (2x − π) + 1.**

And putting it in the general form, we get: **2 sin 2(x − π/2) + 1.**

Then, we can now see that the amplitude is 2, the frequency is 2, the phase is −π/2, that is, the horizontal shift is −π/2 (to the right), and the vertical shift is 1 (upward).

And next, putting it in a graph, we may want to begin with the prototype: **sin x.**

Then, setting first, the frequency to 2, we get: **sin 2x.**

We know that the period is: the basic period (interval) divided by the frequency, and that the basic interval (period) for the sine is 2π. So we can see that the period is: 2π/2 = π.

And thus, we can put the curve with **sin 2x** the way below:

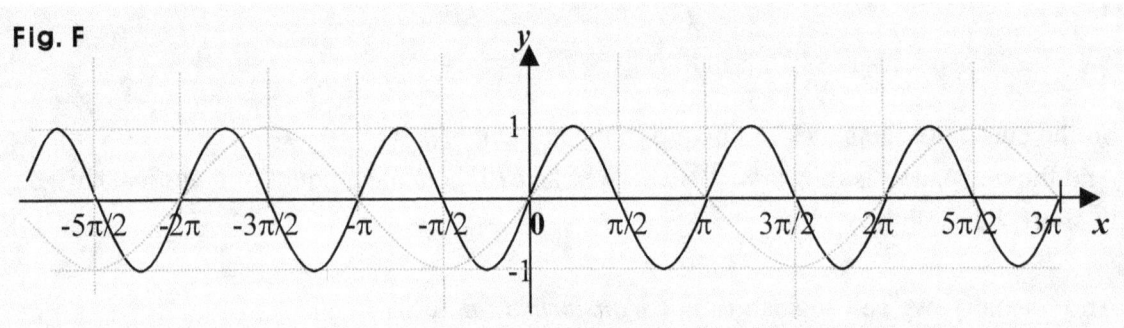

Fig. F

Next, setting the amplitude to 2, we get: **2 sin 2x.**

And the amplitude indicates half the width of the curve. So the half width is 2.

And thus, we can put in a graph, the curve with (**2 sin 2x**) the way below:

Fig. G

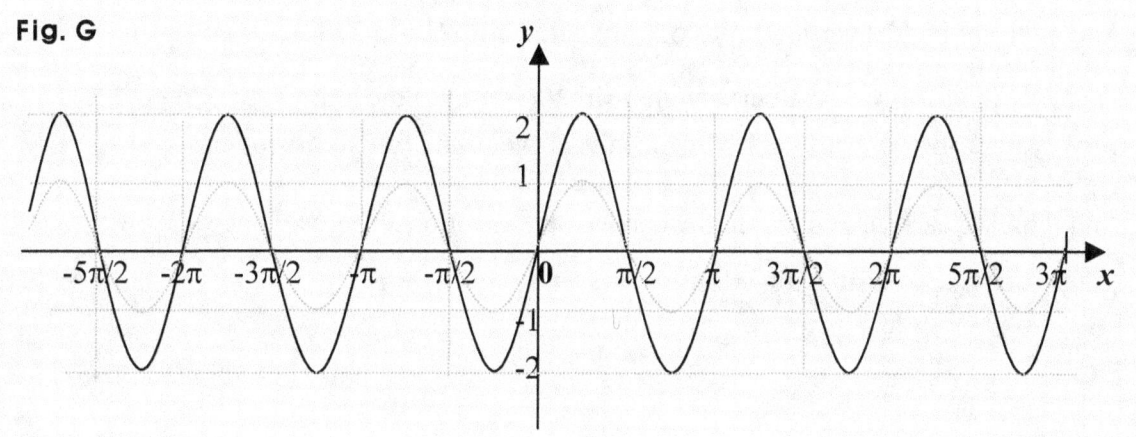

Next, setting the phase to -π/2, we get: **2 sin 2(x – π/2)**.

We know shifting the curve with **sin x** by -π/2 in the direction of the x-axis, we get the curve with **sin (x + π/2)**.

So shifting the curve with **2 sin 2x** by not -π/2 but π/2 in the direction of the x-axis, we get the curve with **2 sin 2(x – π/2)**.

That is, moving the curve with **2sin 2x** in the amount of π/2 to the right, we get the curve with **2 sin 2(x – π/2)**. So we do a horizontal shifting to the right in the amount of π/2.

Note however, if the domain is a set of all real numbers, the curve with **2 sin 2(x – π/2)** is in fact, no other than the curve with **2 sin 2(x + π/2)**.

If however, the phase is not a multiple of **±π/2**, the two curves are different.

And thus, we can put in a graph, the curve with **2 sin 2(x – π/2)** the way below:

Fig. H

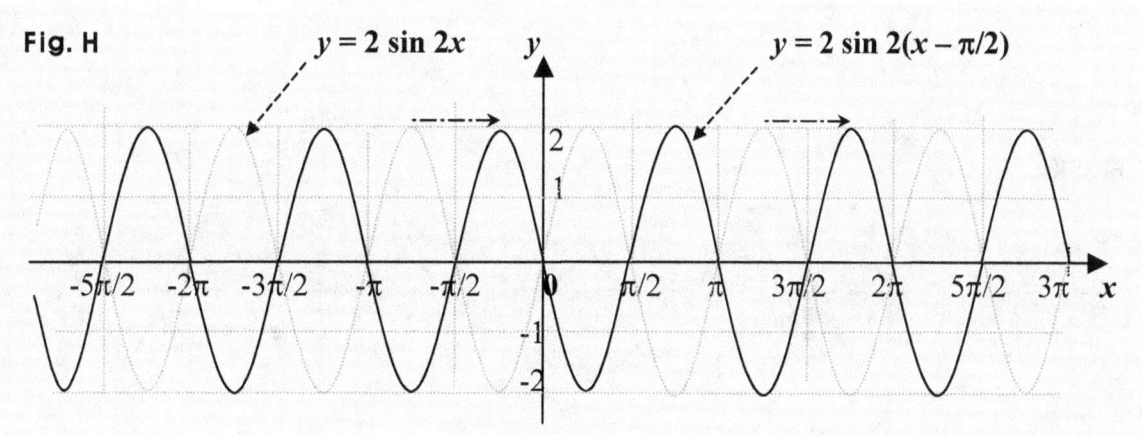

164

And next, setting the vertical shift to 1, we get: **2 sin 2(*x* – π/2) + 1**.

Then, we move upward the curve with **2 sin 2(*x* – π/2)** in the amount of 1 to get the curve with **2 sin 2(*x* – π/2) + 1**, which is the curve of the function **S** given.

And thus, we can put in a graph, the curve of the function **S** the way below:

{Note that: **S(*x*) = 2 sin 2(*x* – π/2) + 1 = -2 sin (-2*x* + π) + 1**.}

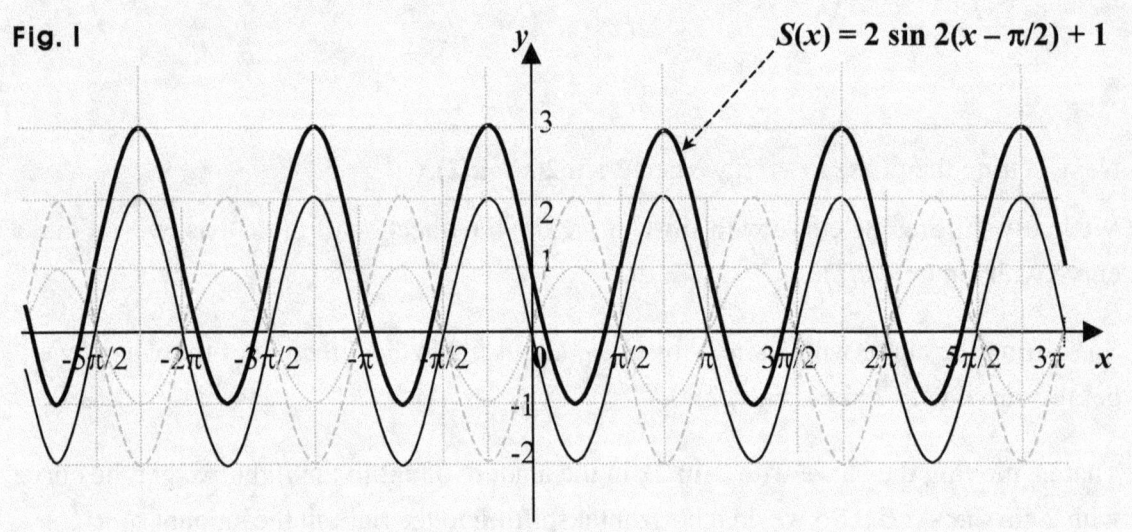

Fig. I

$S(x) = 2 \sin 2(x - \pi/2) + 1$

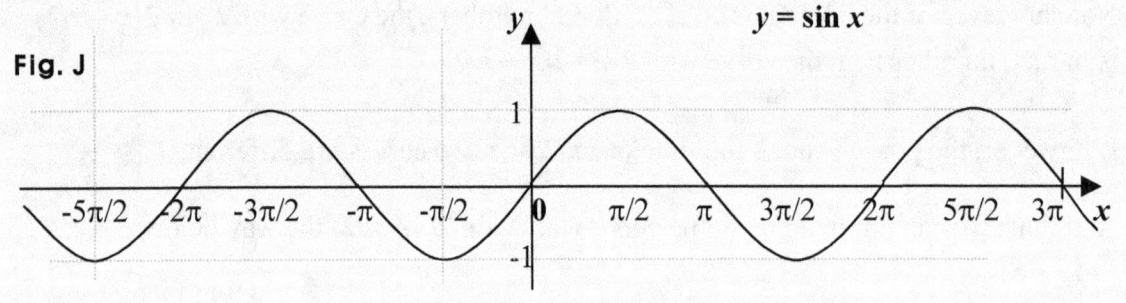

Fig. J

$y = \sin x$

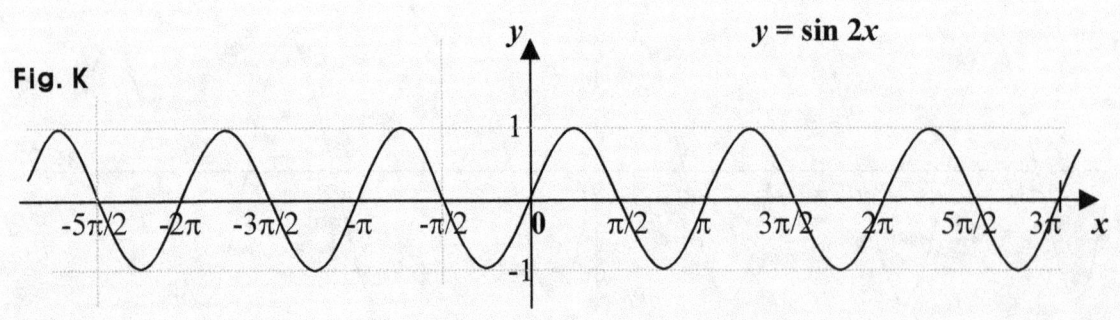

Fig. K

$y = \sin 2x$

Note that $y = S(x) = -2\sin(-2x + \pi) + 1 = 2\sin 2(x - \pi/2) + 1$.

Fig. L

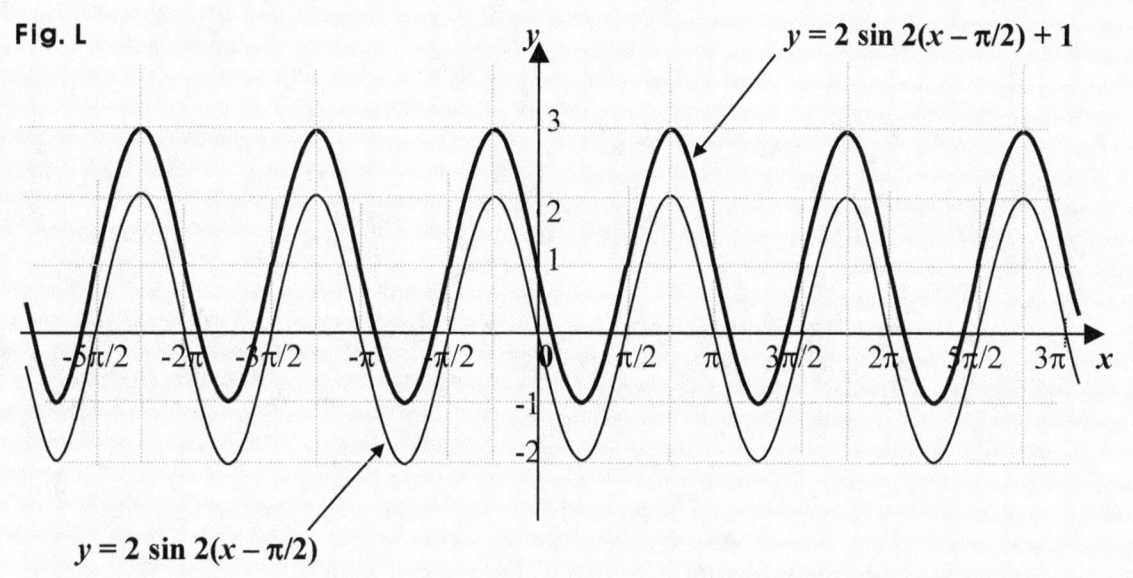

$y = 2 \sin 2(x - \pi/2) + 1$

$y = 2 \sin 2(x - \pi/2)$

www.ingramcontent.com/pod-product-compliance
Lightning Source LLC
Chambersburg PA
CBHW081447170526
45166CB00008B/2345